Silk Reeling

by

S.K. Ananthanarayanan

BIOTECH

2026

Biotech Books

ISBN13 978-81-7622-190-0

Published by : **BIOTECH BOOKS**
4762-63/23, Ansari Road, Darya Ganj,
New Delhi - 110002
Phones : 011 4300 3222
Email : biotechbooks@yahoo.co.in

PRINTED IN INDIA

Preface

'Silk Reeling' is a manual designed to aid the reader in understanding the process of reeling in sericulture. Reeling refers to the extraction of silk from silkworm cocoons and the combination of several silk filaments into one strand. As a consequence, it is one of the sericultural process.

Reeling of silk is a delicate art, and as such must be handled carefully. The silk filaments being extremely fine, we unfortunately cannot reel the strands straight from silkworms, as from spiders. Hence, the reeling of a single fiber of silk uses upto five or ten cocoons, and involves a complicated process by machine in a factory called a filature. The reeled yarn is termed raw silk.

The silk reeling industry covers a wide range of activities included in sericulture—handling of fresh cocoon, cocoon drying and storage, cocoon boiling and reeling, raw silk re-reeling, and finishing, management of water quality and quality, utilisation of by products, maintenance, etc. This book guides the reader through each step of the reeling process, right from the placing of cocoons in hot water to the sorting of skeins of raw silk yarn. A comprehensive and detailed work on this issue, the book should prove to be a highly useful reference guide for silk reeling managers and sericulturists.

Editor

Contents

1

Basics of Silk Culture

Silk is a filament fiber formed from proteins secreted by Bombyx mori, or silkworms. Silkworms are not actually worms, but caterpillars, despite their common name. Humans have practiced silk production, which originated in China, for thousands of years. Highly prized for its softness, insulating properties, and strength, silk is a natural animal product and therefore quite expensive. Making silk requires monitoring and feeding the silkworms constantly, and a great deal of effort results in a surprisingly small amount of thread.

China managed to keep the secret of silk for thousands of years, exporting the rare textile to Europe over trade routes. Eventually, silkworm eggs were smuggled out, and in the 13th century, Western production of silk began in Italy. This by no means brought the cost down, as the extensive amount of work required to make silk remained the same. Thirty thousand eggs can end up eating one ton of mulberry leaves and producing 12 pounds (5.5 kilograms) of silk.

Silk manufacture begins when female silkworms lay their eggs. A single silkworm may produce hundreds of tiny eggs, which are incubated until they hatch into larva. The larva must be kept warm and fed on mulberry leaves frequently throughout their brief lives. In four to six weeks, the larva have reached their maximum size and have mustered enough energy to pupate.

It is the pupation stage that produces silk, as the larva attach to branches and spin a shell around themselves. The shell is unique to insect pupation, formed by two spinnerets on the larva. The fluid they secrete is high in protein and forms a continuous thread, which is repeatedly wound to form a pod to mature in.

Unfortunately for the silkworm pupa, it is maturity that silk producers want to avoid. A small number of the larva are allowed to gestate into adults to carry on the lineage, while the rest are subjected to heat to kill the larva before they can begin to eat through the valuable silk thread. The pods are dipped into hot water to loosen the thread, which is then wound onto wheels. The dead larva are discarded.

Once raw silk has been wound onto wheels, it can be spun into a variety of different types of thread, depending upon the intended use. Crepe is made by twisting multiple strands of silk together in different directions, while tram is made by twisting one to two threads in the same direction. Organzine is formed by twisting multiple threads together, switching direction, and repeating the process until the end of the thread.

Most wearers are familiar with all forms of silk thread, with single threads being used for fine and sheer garments and crepe being used to create textured and wrinkly silks. Organzine is used for warp threads in weaving, and tram creates the weft, or filling. Silk can also be used in knit fabrics.

Silk can be treated to remove roughness or left raw, depending on the demand. Silk takes dye well and is available in a dazzling array of colors from subtle to bright. The textile appears in scarves, sweaters, underwear, shirts, and everything in between. Despite its low yield, production of the sought after textile is rising globally. Many consumers prefer silk for its comfort, insulating qualities, look, and feel.

History of Silk

According to archeological evidence, silk and silk fabric emerged in China at least 5,500 years ago. The cultivation of the silkworm can be traced back to the third century BC. It was said that the demigod Leizu, a legendary figure of prehistoric China, was the first to plant mulberries and raise silkworms. During the Zhou Dynasty (11th century-256BC), a special administration was set up to manage sericulture (silkworm breeding) and silk production. The famous Silk Road to the Middle East and Europe started under Zhang Qian. Under imperial order, he started his diplomatic mission to the West from 138 BC to 126 BC. Gradually, sericulture and silk production techniques spread to other countries. Chinese silk was highly prized among the wealthy of the ancient Roman Empire. Today, Chinese silk still enjoys its reputation for high quality throughout the world.

The business of raising silkworms and unwinding cocoons is now known as silk culture or sericulture. It takes an average of 25-28 days for a silkworm, which is no bigger than an ant, to grow old enough to spin a cocoon. Next, farmers (usually female) will pick them up and place them one by one onto piles of straws. Then each silkworm, with its legs stretched out, will attach itself to the straw and begin to spin.

The next step is unwinding the cocoons, a process that is usually done by "reeling" women. The cocoons are heated to kill the pupae, which must be done at the right time; otherwise, the pupas are bound to turn into moths. (Moths make a hole in the cocoon, an event that makes reeling useless.)

To unwind the cocoons, first they are put into a basin filled with hot water. Then the reeling women find the loose end of the cocoons, and then twist them. Afterwards, the women carry the cocoons to a small wheel for unwinding. At last, two workers measure them into a certain length and twist them into so-called "raw" silk, which then are dyed and woven into cloth. An interesting fact is that about 1,000 meters of can be unwound from one cocoon, while 111 cocoons are needed for a man's tie, and 630 cocoons are needed for a woman's blouse.

Making of Silk in China

The making of silk generally refers to the process of dividing raw silk from cocoons into strands horizontally and vertically, before weaving them together into pieces of fabric. The actual manufacture processes of various silks vary, but can be generally categorised into two types: sheng zhi and shu zi.

In the *sheng zhi* process, weavers weave the raw silk into fabrics first, and then scour (clean) and bleach the fabrics. This process, which has lower costs and a shorter process, is currently the major way of making silk.

In the *shu zhi* method, weavers scour and bleach the longitude and latitude silks from cocoons first before actually weaving them. The woven products no longer need further processing and can be directly used. The method is usually used to produce advanced silk fabrics like brocade.

Before the silk is woven, a lot preparations need to

be done beforehand, like soaking the raw silk to soften the product. Meanwhile, as silk is very apt to absorb moisture, to make the silk damp-proof is very important before the weaving.

In terms of the silk pattern, the weaving methods can be generally divided into the common and jacquard methods. The former refers to the flat silk fabrics that have no weaved patterns, while the latter refers to the fabrics that are usually done by a jacquard loom (a loom that is mechanised to weave specific patterns). When the silk fabrics are ready, the next step is the dyeing process (which is crucial in the whole procedure of making colorful and beautiful silk). With the dyeing technologies, the raw silk can be turned into flawless silk with patterns and colors to people's desire.

In ancient China, once the cloth had been weaved, embroidery was used to give the cloth its delicate, often brilliant patterns. The Four Renowned Embroideries of China were regional in their origin: Su embroidery originated from East China's Jiangsu Province; Yue embroidery originated from South China's Guangdong Province; Xiang embroidery originated from Central China's Hunan Province; and Shu embroidery originated from Southwest China's Sichuan Province.

Silk Kingdom

Silk, as a symbol of ancient Chinese culture, has not only weaved an excellent picture in the nation's civilisation history, but also has made indelible contributions for the advancement of human beings. For thousands of years, Chinese silk has been known for its superior quality, exquisite patterns, and rich cultural connotations.

Several thousand years ago, when the silk trade first

reached Europe via the Silk Road, it brought with it not only gorgeous silk apparel and decorative items, but also the ancient and resplendent culture of the Far East. From then on, silk was regarded as the emissary and symbol of Eastern civilisation. The earliest silk article discovered to date is approximately 4700 years old, unearthed from a tomb dating from China's Liangzhu Culture (c. 3300-2200 BC).

According to an ancient Chinese legend, the Silkworm Goddess appeared to the Yellow Emperor, the legendary ancestor of the Chinese people, after he had vanquished his adversary Chi You. She presented him with silk fibers spun from her own mouth as a sign of respect. The Yellow Emperor ordered the fibers woven into cloth and made into silk apparel, which he found exceedingly soft and comfortable.

Lei Zu, the Yellow Emperor's wife, searched until she found some caterpillars capable of spinning silk fibers from their mouth. She raised these silkworms by feeding them mulberry leaves she picked herself. Later generations came to worship Lei Zu as the Silkworm Goddess, and the Yellow Emperor as the God of Weaving.

Sericulture, including cultivating the mulberry plant, raising silkworms, and producing silk fabric, has been an essential form of labor in China throughout the millennia, as China is the birthplace of sericulture. Raising silkworms and reeling the silk from their cocoons was ancient China's greatest achievement in the use of natural fibers.

As long ago as the Neolithic Age (c. 12,000-2000 BC), the Chinese ancestors had invented flat-weaving and figured-weaving techniques, and were tinting cloth using natural (red pigment) vermilion dye. With improvements in looming devices and printing and dyeing methods, more varieties of silk were developed and a comprehensive system of cloth dying evolved. China possessed the most advanced silk dying and weaving techniques of the ancient world.

Silk Road

It is well known that silk is discovered in China as one of the best materials for clothing - it has a look and feeling of richness that no other materials can match. However, very few people know when or where or how it is discovered. Actually, it could date back to the 30th Century BC when Huang Di (Yellow Emperor) came into power. There are many legends about the discovery of silk; some of them are both romantic and mysterious.

Legend has it that once there lived a father with his daughter, they had a magic horse, which could not only fly in the sky but also understand human language. One day, the father went out on business and did not come back for quite some time. The daughter made him a promise: If the horse could find her father, she would marry him. Finally her father came back with the horse, but he was shocked at his daughter's promise.

Unwilling to let his daughter marry a horse, he killed the innocent horse. And then miracle happened! The horse's skin carried the girl flying away. They flew and flew, at last, they stopped on a tree, and the moment the girl touched the tree, she turned into a silkworm. Everyday, she spit long and thin silks. The silks just represented her feeling of missing him.

Another less romantic but more convincing explanation is that some ancient Chinese women found this wonderful silk by chance. When they were picking up fruits from the trees, they found a special kind of fruit, white but too hard to eat, so they boiled the fruit in hot water but they still could hardly eat it. At last, they lost their patience and began to beat them with big sticks. In this way, silks and silkworms were discovered. And the white hard fruit is a cocoon!

The business of raising silkworms and unwinding cocoons is now known as silk culture or sericulture. It takes an average of 25-28 days for a silkworm, which is no bigger than an ant, to grow old enough to spin cocoon. Then the women farmers will pick them up one by one to piles of straws, then the silkworm will attach itself to the straw, with its legs to the outside and begin to spin.

The next step is unwinding the cocoons; it is done by reeling girls. The cocoons are heated to kill the pupae, this must be done at the right time, otherwise, the pupas are bound to turn into moths, and moths will make a hole in the cocoons, which will be useless for reeling. To unwind the cocoons, first put them in a basin filled with hot water, find the loose end of the cocoon, and then twist them, carry then to a small wheel, thus the cocoons will be unwound. At last, two workers measure them into a certain length, twist them, they are called raw silk, then they are dyed and woven into cloth.

An interesting fact is that we can unwind about 1,000 meters long silk from one cocoon, while 111 cocoons are needed for a man's tie, and 630 cocoons are needed for a woman's blouse.

Chinese people developed new way by using silk to make clothes since the discovery of silk. This kind of clothes became popular soon. At that time, China's technology was developing fast. Emperor Wu Di of western Han Dynasty decided to develop trade with other countries. To build a road becomes priority to trade silk. For nearly 60 years of war, the world famous ancient Silk Road was built up at cost of many losses of life and treasures. It started from Chang'an (now Xi'an), across Middle Asia, South Asia and West Asia. Many countries of Asia and Europe were connected.

From then on, Chinese silk, along with many other Chinese inventions, were passed to Europe. Romans, especially women, were crazy for Chinese silk. Before that, Romans used to make clothes with linen cloth, animal skin and wool fabric. Now they all turned to silk. It was a symbol of wealth and high social status for them to wear silk clothes. One day, an Indian monk came to visit the Emperor. This monk had been living in China for several years and knew the method of raising silkworms. The Emperor promised a high profit of the monk, the monk hid several cocoons in his cane and took it to Rome. Then, the technology of raising silkworms spread out.

Thousands of years have passed since China first discovered silkworms. Nowadays, silk, in some sense, is still some kind of luxury. Some countries are trying some new ways to make silk without silkworms. Hopefully, they can be successful. But whatever the result, nobody should forget that silk was, still is, and will always be a priceless treasure.

Silk Production: China to Rest of the World

In spite of their secrecy, however, the Chinese were destined to lose their monopoly on silk production. Sericulture reached Korea around 200 BC, when waves of Chinese immigrants arrived there. Silk reached the West through a number of different channels. Shortly after AD

300, sericulture traveled westward and the cultivation of the silkworm was established in India.

It is also said that in AD 440, a prince of Khotan (today's Hetian)—a kingdom on the rim of Taklamakan desert — courted and won a Chinese princess. The princess smuggled out silkworm eggs by hiding them in her voluminous hairpiece. This was scant solace to the silk-hungry people of the West, for Khotan kept the secret too. Why share it with the westerners and kill a good market?

Then around AD 550, two Nestorian monks appeared at the Byzantine Emperor Justinian's court with silkworm eggs hid in their hollow bamboo staves. Under their supervision the eggs hatched into worms, and the worms spun cocoons. Byzantium was in the silk business at last. The Byzantine church and state created imperial workshops, monopolizing production and keeping the secret to themselves. This allowed a silk industry to be established in the Middle East, undercutting the market for ordinary-grade Chinese silk.

However high-quality silk textiles, woven in China especially for the Middle Eastern market, continued to bring high prices in the West, and trade along the Silk Road therefore continued as before. By the sixth century the Persians, too, had mastered the art of silk weaving, developing their own rich patterns and techniques. It was only in the 13th century—the time of the Second Crusades—that Italy began silk production with the introduction of 2000 skilled silk weavers from Constantinople. Eventually silk production became widespread in Europe.

Global Silk Trade

Silk became a precious commodity highly sought by other countries at a very early time, and it is believed that the silk trade was actually started before the Silk Road was officially opened in the second century BC. An Egyptian female mummy with silk has been discovered in the village of Deir el Medina near Thebes and the Valley of the Kings, dated 1070 BC, which is probably the earliest evidence of the silk trade. During the second century BC, the Chinese emperor, Han Wu Di's ambassadors traveled as far west as Persia and Mesopotamia, bearing gifts including silks. A Han embassy reached Baghdad in AD 97, and important finds of Han silks have been made along the Silk Road. One of the most dramatic finds of Tang silks along the Silk Road was made in 1907 by Aurel Stein. Some time around 1015, Buddhist monks, possibly alarmed by the threat of invasion by a Tibetan people, the Tanguts, sealed more than ten thousand manuscripts and silk paintings, silk banners, and textiles into a room at the Caves of the Thousand Buddhas near Dunhuang, a station on the Silk Road in north-west Gansu.

From about the fourth century BC, the Greeks and Romans began talking of Seres, the Kingdom of Silk. Some historians believe the first Romans to set eyes upon the fabulous fabric were the legions of Marcus Licinius Crassus, Governor of Syria. At the fateful battle of Carrhae near the Euphrates River in 53 BC, the soldiers were so startled by the bright silken banners of the Parthian troops that they fled in panic. Within decades Chinese silks became widely worn by the rich and noble families of Rome. The Roman Emperor Heliogabalus (AD 218 - 222) wore nothing but silk. By 380 AD, Marcellinus Ammianus reported, "The use of silk which was once confined to the nobility has now spread to all classes without distinction, even to the lowest." The craving of silk continued to increase over the centuries. The price of silk was very hight in Rome. The best Chinese bark (a particular kind of silk) cost as much as 300 denarii (a Roman soldier's salary for an entire year!). Many sources quote that Roman citizens' demand for imported silks was so great as to be damaging to the Roman economy.

Silk was even beginning to have a civilizing effect on the barbarians. In 408 AD when Alaric, a Goth, besieged Rome, his price for sparing the city included 5000 pounds of gold, 3000 pounds of pepper, 30,000 pounds of silver and 4000 tunics of silk.

Silk Production Today

Silk production today is a blend of ancient techniques and modern innovations. The first stage of silk production is hatching the silkworm eggs, which have been previously examined and shown to be free from disease. Larvae are then fed cut-up mulberry leaves and after the fourth molt climb a twig placed near them and spin their silken cocoons. The silk is a continuous-filament fiber consisting of fibroin protein secreted from two salivary glands in the head of each larvae, and a gum called sericin, which cements the two filaments together. Pupae within cocoons are killed by steam or fumigation to prevent adult emergence, which would cut and tangle the silk filaments. Cocoons are latter softened in hot water to remove the sericin, thus freeing silk filaments for reeling. Single filaments are drawn from cocoons in water bowls and combined to form yarn. This yarn is drawn under tension through several guides and eventually wound onto reels. The yarn is dried, packed according to quality, and is now raw silk ready for marketing.

World silk production has approximately doubled during the last 30 years in spite of man-made fibers replacing silk for some uses. China and Japan during this period have been the two main producers, together manufacturing more than 50% of the world production each year. China during the late 1970's drastically increased its silk production and became the world's leading producer of silk. The 1970's were a period of tumultuous political and social upheaval in China, resulting in various economic reforms. Undoubtedly, these reforms are partially responsible for China's increased silk production. Thus the country that first developed sericulture approximately 4,700 years ago has again become the world's main producer of silk.

Feeding of Silkworms

Silkworms go through four stages of development, as do most insects: egg, larva, pupa and adult. The adult (imago) stage is the silkworm moth. The larva is the silkworm caterpillar. Since the silkworm grows so much, it must shed its skin four times while it is growing. These stages-within-a-stage are called instars. Silkworms only eat fresh mulberry leaves (or artificial food). Mulberries are deciduous and only have leaves in the late spring, summer and early autumn.

Mulberry Cultivation

Mulberry can be grown under various climatic conditions. Its growth depends on many climatic conditions such as temperature, humidity, rainfall etc. The soil should be deep, fertile, well drained, clay loam to loam, with good moisture holding capacity. Slightly acidic (6.2-6.8 pH) soil free from injurious salts is ideal for good growth of mulberry.

Preparation of field

The field is levelled and ploughed deeply before the onset of monsoon. FYM may be applied @ 10 t/ha for the rainfed crop and 20 t/ha for the irrigated crop during land preparation. Following are two important method of planting and spacing:

- Pit system (rainfed crop): Spacing 75 x 75 cm (pit size 30 x 30 x 30 cm)
- Row system (irrigated crop): Spacing 60 x 60 cm (ridges and furrows)

Planting material

The variety K_2 gives higher yield and better quality leaves. Cuttings must be prepared from shoots of proper maturity (6-8 months) and thickness with well-developed buds. Cuttings of 7-10 cm length and pencil thickness with 3 or 4 active buds are ideal.

For irrigated crop, two cuttings should be planted at each spot along the margin of the ridge. For rainfed crop, three cuttings are to be planted per pit in a triangular manner with a distance of 15 cm, keeping only one bud exposed.

After 8 months of planting, 50 kg each of N, P_2O_5 and K_2O per ha should be applied after weeding. First harvest can be taken six months after planting by leaf picking. Second dose of 50 kg N per ha should be applied 8 weeks after the first leaf harvest. Two more crops can be taken at an interval of 3 months, by leaf picking.

Manuring

For rain fed crop apply FYM @ 10 t/ha as a basal dose and topdress every year at the time of annual pruning. Fertilisers are applied @ 130:65:65 kg/ha of $N:P_2O_5:K_2O$ in two split doses. For irrigated crop, FYM is given @ 20 t/ha as basal dose. Fertilisers are applied @ 300:120:120 kg/ha of N : P_2O_5 : K_2O in five split doses.

Pruning

For rainfed crop, bottom pruning is done in May-June. Two top clippings in August/September and December/January are also practised. Middle pruning is done in October/November. For irrigated crop, bottom pruning at 15-30 cm height in May, two top clippings in August and December and two middle pruning at 60 cm height in October and February/March are practised.

Pests

Tussock Caterpillars: Larvae eat the leaves of the mulberry plant. Their incidence is frequent during March to August. Collection and destruction of egg masses and spraying 1% DDVP arc cffective. Waiting period is 3 days.

Jassids (Empoasca flarescens): Greenish hoppers feed on the underside of the leaf, sucking sap causing hopper burn. Spraying 0.1% dimethoate is effective. Waiting period is 10 days.

Thrips: These are frequent during summer season. Attack is severe in rainfed gardens. Spraying 0.02% DDVP is effective. Waiting period is 3 days.

Mealy Bugs (Maconelliococcus Hirsutus): It causes 'tukra disease'. The affected leaves show curling and stunted growth at the growing point. Application of methyl demeton (0.05%) is effective. Waiting period is 15 days.

Scale Insect: When attack is severe, branches dry and become yellow. Spraying lime sluphur solution is effective.

Leaf Eating Caterpillar (Diacrisia Obliqua): Appears frequently between November and January. Collection and destruction of egg masses, deep ploughing and flood irrigation to kill the pupae and application of 0.2% DDVP on the leaves can prevent the attack.

Diseases

Root Knot Disease (Meloidogyne Incognita). Common in sandy loam type of soil under irrigated conditions. Controlled by applying neem oil cake @ of 400 kg per ha per year in four equal split doses.

Powdery Mildew (Phyllactina Corylea). It is more common during November-February. White powdery patches appear on the lower side of the leaves. Can be controlled by spraying dinocap 0.2%.

Leaf Rust (Ceratelicum fici). The attacked portion of the leaves has whitish brown pustule on both sides and is deformed and also not nutritive. Infection is more in November-February. This can be controlled by spraying carbendazim 0.1% or tridemorph 0.1%

Leaf Spot (Cercospora moricola). Diseased leaves have a number of circular or irregular brownish black spots of varying size. Infection is more common in rainy season. This can be controlled by spraying 0.1% of carbendazim.

Silkworm Rearing

Requirements for Silkworm Rearing

— Good quality mulberry leaves

— Rearing house of approximately 20 m^2 for 100 dfls (disease free layings), with good ventilation, mild temperature (24-28°C) and humidity (65-85%).

— Rearing equipments like chawki stand (one), wooden trays (10), rearing racks (5), chopping board (one) and knife, wooden/bamboo rearing trays (50), chandrika/netrika (mountage) (40), leaf chamber, feeding stands, ant wells, rocker sprayer, wet and dry bulb thermometer and materials like formaldehyde/bleaching powder, paraffin paper, cleaning nets, foam rubber strips, and RKO powder are required.

Techniques of Rearing

Disinfect the rearing house and equipments to prevent silkworm disease, two-three days before rearing. First, wash the rearing house and the equipments with 2% bleaching powder. Then spray the room and equipments with 5% bleaching powder or 2% formaldehyde. Close the houses for 24 hours for the fumes to get diffused.

First incubate the dfls (egg card) at a temperature of 24-26°C and RH 75-80%, one day prior to hatching (blue egg stage); cover the eggs with black paper (black boxing). Next day morning, open it and expose to diffused sunlight. As the larvae emerge out, fresh tender leaves collected from the plant are chopped into 0.5 x 0.5 mm size and sprinkled over the hatched larvae. After half an hour, transfer the larvae to the paraffin paper spread in the chawky trays (wooden trays) using fine brush. Provide wet foam strips around and prepare a compact bed. Give another feeding in the bed. Cover with paraffin paper and stack the trays one over the other on the stand. Up to 20 laying can be brushed in a tray of 90 x 60 cm.

Feeding Schedule

At the end of each instar, larvae stop feeding and cast off old skin in 18-30 hours. When the worms set for moulting, paraffin paper should be removed and spread on the bed to dry up. If there are more feeding worms, a light and thin feeding may be given. All the worms settle in 6-8 hours. During moulting, worms should not be disturbed and full ventilation should be provided. Feeding is resumed when 90% of worms have moulted. RKO powder is dusted over the worms 30 minutes before feeding. After two consecutive feedings, the larvae with the net are transferred to a new tray. Mature larvae stop feeding and prepare themselves for spinning.

Its body becomes translucent, shrinks in length and constrictions appear on fourth and fifth segments. They move towards the periphery of the trays. Such worms are picked and transferred to Chandrika/Netrakae. About 1000 worms (400-450 larvae/m^2) can be mounted in a mountage. Mount the entire larvae within a maximum period of 48 hours and provide sufficient ventilation during spinning. Cocoon should be harvested on the fifth and sixth day after mounting. In rainy and cold seasons, it should be delayed for one more day.

The cocoons are collected from Chandrika and transported in light gunny bags to cocoon market. The cocoon should be marketed immediately after harvest, so as to avoid adult emergence. Under average conditions, 100 dfls of bivoltine will yield 40-60 kg cocoons and cross breed will yield 30-50 kg cocoons.

Diseases

Pebrine

It is the most destructive disease caused by protozoa, Nosema bombyscis. The worms become very dull and they have poor appetite, irregular moulting and the skin becomes wrinkled.

Flacherie

Caused by bacteria, promoted by high temperature, high humidity and ventilation, bad leaf quality, over feeding and low alkalinity of the gut. Digestive and circulatory systems are damaged and the symptoms are loss of appetite and diarrhoea.

Grasserie

Mostly seen in riping larvae. Caused by Borrlina virus. Infection is induced by extreme low and high temperature. Swelling of the inter-segmental region, shining skin, rupture of body wall, oozing of body fluid and endless crawling are symptoms. Such worms do not moult and spin.

Mascardine

The fungi Beauveria bassiana, Spicaria prasina and Isaria farinosa are causal agents. The infected larvae lose appetite. Specks of oozing oily substance without any clear-cut margins appear on the skin. Body generally hardens and becomes stiff.

Prevention and Control of Diseases

- Disinfect the rearing room and equipment before rearing.
- Use only disease free layings from authorised agencies.
- Dip the egg cards in 2% formalin solution for 20 minutes before incubation.
- Collect undersised larvae and destroy regularly by burning or burrowing in soil.
- Feed good quality leaves of correct stages.

— Avoid over feeding and under feeding
— Clean the bed every day and burn the infested litter.
— Use RKO powder at every moulting before resumption of feeding.
— Maintain humidity only to the desired level.

2

History of Silk Reeling

The 20th century eclipse of Chinese technological leadership in silk production by Japan in fact had its early modern historical precedent. Although it is beyond dispute that the prototype of the foot-treadle silkreeling machine in Southern Europe was fundamentally Chinese, or more specifically, Lower Yangzi in origin, the French and Italian machines, aided by a series of subtle but crucial technological innovations, had risen to world leadership by the late-17th century.

These innovations came to East Asia around the mid-19th century as a bundle consisting of four new features. The first was the adoption of rigid axle and cogwheel mechanism to transmit the power to move the reel in place of the driving-belt it had borrowed from China. The second was the design of an additional twisting mechanism to cross silk threads dry. The first mechanism stabilises the reeling movement, while the second innovation greatly enhances the cohesiveness, evenness and uniformity of the silk thread, features that are absolutely essential for high quality raw silk. The third was the use of a centralised steam boiler that provided continuous and even heating for the reeling basins of cocoons. The fourth, which came the last, was mechanisation. However, the use of steam power had not been crucial early on for the reeling of delicate silk threads - even in Europe, fully mechanised silk-reeling was mostly a 20th century phenomenon. Therefore, the major distinction between hand and machine-reeling in the 19th century East Asia was the use of centralised steam heating.

Entrepreneurs in East Asia had selectively adopted various features of this bundle to suit their needs. For example, the traditional silk-reelers encountered little difficulties to insert the first two features of European reeling technology - cogwheels and crossing systems - into their household based production. However, the adoption of the third and fourth technical innovations required the use of a centralised location, a first step towards the rise of modern factories. But even there, as will be shown below, various technological and organisational compromise had evolved in different periods that the distinction between household and factory production could be blurry at times in the 19th century East Asia.

In comparison with other modern industries, silk-reeling factories remained relatively small in scale, low in capital intensity and dispersed in location. The evenness and uniformity of the machine-reeled raw silk found favour with the weavers in Europe and particularly the U.S, who were growing increasingly in scale and capital intensity. It commanded a price premium over the hand-reeled silk in the international market. Throughout the 19th and early 20th century, average price for machine-reeled raw silk were about 40% higher than hand-reeled and 25% higher than the so-called improved hand-reeled raw silk, which adopted an additional process of hand-re-reeling.

Yet despite the price premium, the diffusion of machine-reeled silk did not lead to the immediate displacement of hand-reeled silk in exports. Table 1 shows that exports of hand-reeled silk in China and Japan rivalled that of the machine-reeled before 1900. Even after the 20th century that machined reeled began to dominate only in Japan and Guangdong but not the Lower Yangzi. Guangdong, partly due to its relatively small level of hand-reeled silk exports, had the highest share of exports in machine-reeled silk, followed closely by Japan. On the other hand, machine-reeled silk exports from the Lower Yangzi, the world's longtime leading high quality silk producing region, were so puny that they were not even separately counted in the 1893 trade statistics. Even towards the 1920s, machine-reeled silk counted only about half of the total exports from Shanghai.

Table 1. Raw Silk Exports of China and Japan by Types (in Metric Tons)

Years	Hand-Reeled Silk			Machine-Reeled Silk		
	Lower Yangzi	Guangdong	Japan	Lower Yangzi	Guangdong	Japan
1880-90	2370	345	1074	—	360	728
1890-1900	2576	160	1010	484	1349	2358
1900-10	1715	93	549	834	2097	5344
1910-20	1320	127	153	1609	2206	11825
1920-30	742	37	164	2306	2877	23989

To analyse the persistence of hand-reeled silk and the differential rate of diffusion of machine-reeling across these three regions, we can construct a simple partial equilibrium model. As demand side of raw silk is set in the international market, we can treat the prices of machine-reeled and handreeled silk, denoted as PM and PH respectively, as parameters. For the supply side, we posit that the output of machine and hand-reeling silk output (Y_M and Y_H– respectively) are produced by two constant returns to scale technologies, each using three inputs: Labour (L), Reeling Machines (K) and Cocoons (Q). We assume that the total quantities of cocoons produced for each year are fixed and shared by machine and hand-reelers. That is $Q = Q_M - + Q_H$. It is also reasonable to assume that machine-reeled technology has higher total factor productivity (A_M) than that of the hand-reeled (A_H), that is $A_H = \alpha A_M$ where $0<\alpha<1$. We can write them out as:

$$Y_M = A_M\ F(L, K, Q_M);$$
$$Y_H = \alpha A_M\ F(L, K, Q_H).$$

The corresponding cost equations for these production functions are:

$$C_M = w_L + (1+r)(P_{KM}K_M + P_Q Q_M) + TY_M;$$
$$C_H = w_L + (P_{KH}K_H + P_Q Q_H) + \phi TY_H;$$

where w, r, P_Q, PK_M, P_{KH}, T are wage rates, interest rates, cocoon prices, prices of machine-reel tools and hand-reel tools, transaction costs per unit of output respectively. For machine-reeling sector, we use the (1+r) term as we assume purchase of cocoons and reeling machines are externally financed, while these purchases are self-financed for hand-reeling. As machinereeling tools are more expensive than that for hand-reeling, we have $P_{KM} > P_{KH}$. If ϕ is a coefficient between 0 and 1, which means per unit output transaction cost is higher for machined reeled silk than that of the hand-reeled.

The two cost identities are standard except for the addition of the per unit transaction cost T. Transaction cost here denotes two broad categories of costs facing a silk-reeler. The first and the most straightforward is costs incurred in various market transactions that involved the marketing and moving of silk, the recruitment of labour, the procurement of inputs, particularly cocoons. The second category pertains to the public and government sector, in which the external legal environment, contract enforcement mechanism, securities of properties and government taxation policies are all likely to impact transaction costs. It is quite reasonable to expect that a machine-reeled firm, with larger capital investment concentrated in a centralised location, would likely incur higher transaction cost per unit of output than a hand-reeling unit in both categories. Combine the output price, production function and cost identities, we can derive the following two profit functions machine and hand-reeled silk respectively. The intuition of the equation is that although machine-reeled silk can be produced with higher total factor productivity and fetches a higher price than that of the hand-reeled silk, it also requires more expensive machines, additional interest costs and higher per unit transaction cost.

$$\Pi_M = (P_M A_M - T)Y_M - [w_L + (1+r)(P_{KM}K_M + P_Q Q_M)];$$

$$\Pi_H = (P_H A_H - \phi_T)Y_H - [w_L + (P_{KH}K_H + P_Q Q_H)];$$

With CRS technology, perfect competition and relatively inelastic cocoon supply, profit maximisation with respect to inputs could yield a unique equilibrium ratio of machine-reeled silk to hand-reeled (Y^*_M / Y^*_H) at which the optimal per unit output profit for machine and hand-reeling silk are equal given certain parameters restrictions. Comparative statistics would show that optimal output of machine-reeled silk would rise relative to that of hand-reeled silk if there is an increase in P_M, A_M, w, or a decline in r and P_{KM}.

Note that a decline in T, the per unit transaction cost, will lead to an increase in both Y^*_M, and Y^*_H, but will increase Y^*_M proportionately more than Y^*_H (because $\phi < 1$) so that in the end, Y^*_M / Y^*_H will rise. This model aims to capture several important aspects of the transfer of European reeling technology and organisation in traditional East Asia.

Despite the relatively high profitability and higher productivity of the machine-reeling technology, the viable transplantation and diffusion of the new European technology required substantial learning efforts, and more importantly, vigorous adaptation of a relatively expensive modern machine to the labour-abundant, capital scarce economy of 19th century East Asia. Moreover, to produce in a centralised location and later modern factory compelled substantial reorganisation and realignment on the existing social and economic structures of traditional East Asia.

Being fundamentally ruralbased, the success of modern silk reeling industries hinges on the existing capacity of grass-root rural communities and agrarian structures to respond to these challenges and opportunities. As machine-reeling was often in direct competition to the traditional handicraft interests, our analysis of its differential patterns of diffusion would have to be placed in the larger political economy background of these two nations at this critical junction which saw unprecedented political and economics shocks.

Advent of European Technology

Thanks to the detailed archival records of East Asia's most powerful trading firm, Jardine, Matheson & Co. we are able to learn the dramatic details of the first European silk reeling technology transfer to East Asia. In 1859, Jardine employed a British subject John Major, who, with 15 years of operating silk filature in Naples, Italy, was eager to construct such a factory in China employing "Italian and French methods." In January 1860, factory construction began in Shanghai, a treaty port opened to Western trade since 1843. To construct a modern style factory complete with one hundred mechanised, steam-power driven machines and warehouse was no simple task in the then largely primitive "frontier" environment of Shanghai. The construction was finally completed after a protracted and trouble-ridden seventeen months.

Industrial capital and factory construction proved to be the easier part of this modern factory, named as Ewo Silk Filature. To develop a stable corps of labour force with skill and discipline, in 1861 Major managed to recruit four female French skilled reelers directly from Lyon on a five-year contract basis. Despite some initial set-backs and with all the French girls having eloped with French soldiers within the next two years, a reasonably skilled Chinese female labour force, consisting mostly of refugees from the Taiping rebellion ravaged Nanjing, turned out products of quality comparable to the European standard. The size of the factory also doubled to include an additional 100 reeling basins in 1863.

But the real problem seemed to lie outside the factory or even Shanghai. The treaty port status of Shanghai provided Westerners and their business extraterritorial protection. But the Ewo filature was set up with the presumption that Shanghai's hinterland, the Lower Yangzi delta, could be relied on as a supply base of cocoons, the raw materials for mechanised silk-reeling. This presumption, however, turned out to be fatal. Worldwide, most reeling factories located near cocoon producing regions as transporting finished raw silk is far

Finishing

Conventional method of finishing includes calendaring, weighing, scrapping, starch and glue finish, etc. to impart/improve certain characteristics like drape, fall, handle, feel, stiffness, weight, etc. Recently special finishes like anticrease and flame retardant finishes are attempted on silk. Now you have observed different processes in post cocoon technology from cocoon selection to yarn finishing. Answer the following questions.

— Quantity of raw silk production (in the visited reeling unit) per month/annum

— Quantity of cocoons used to get one kg of raw silk (Renditta)

— Cost of production per kg of raw silk/yarn

— Quantity of silk used to produce one silk saree (its length and breadth)

— No. of cocoons needed to produce one silk saree

— Time taken to weave one silk saree

— Quantity of silk jarigai used/saree

— Use of floss obtained at the time of cooking and brushing and its cost/kg.

— By-products obtained from the reeling unit.

Denier:

It is a unit expressed to denote the thickness of silk filament. It is the weight of 9000 m of silk filament expressed as gram. Filament denier decreases from outside to inside gradually.

Layers of Shell:

— Outer layer / floss / Blaze: Non-reelabale waste

was able to buy all the high-quality cocoons he required. But in another trip to the rural Lower Yangzi a couple of months later, he died of heat apoplexy. Soon after that, Jardine decided to close down the Ewo filature as soon as the current inventory was unwound.

By the May of 1870, the Ewo, East Asia's first mechanised silk reeling plant came to an end after ten years of operation. The reality behind the rise and fall of the Ewo filature, which seemed so intricately bound with the tough-minded and single-willed personality of John Major, was that the factory had been incurring a constant loss throughout its years of operation. It is clear that the fixed cost of setting up an entire infrastructure to support a single factory was simply too high. The next ten years after the closing of Ewo went by with no more new attempts at mechanised silk-reeling in Shanghai.

By the late 1870s, a new technological innovation of drying and storing fresh cocoons at the site of purchase was discovered. This began to stir up new interest in establishing filatures. In the early 1880s, an American trading firm, Russell & Co., and two other Western trading firms constructed new filatures respectively. By 1882 the Jardine's also came back to business by erecting a new Ewo filature factory in 1882. An important shift in strategy of the new Ewo filature was their enlisting the Chinese cocoon merchants as subscribers to stocks and board members.

Most noticeably in 1882, a Chinese comprador of Western silk trading firms, Huang Zhuoqin, started the first Chinese owned filature, marking the beginning of a trend where the entire Shanghai silk industry was to come into the hands of Chinese capital. This wave of construction increased the nascent Shanghai silk industry to about 800 reeling basins.

These developments soon alarmed the traditional silk merchants in the Lower Yangzi. Through petitions and publications, they mobilised the local governors of Shanghai, and the provinces of Jiangsu and Zhejiang to issue a closure order for the two Western silk reeling factories in Shanghai in September 1882. Their order, although rejected by the local British and American authorities that had jurisdiction over their areas of international settlement in Shanghai, sent the stock prices of these two factories plunging and caused the closure of their rural cocoon station.

In 1883, Li Hongzhang, China's most powerful bureaucrat at the time, made a compromise and allowed the existence of these two Western silk reeling factories in Shanghai. Subsequently, other Chinese owned filatures in Shanghai were also ordered to register with the authority. Part of the consideration for the compromise was the government's levying of a transit tax, known as the likin tax, on cocoons the filatures purchased in the rural area. The opening of cocoon stations, like the silk hongs, would also be subject to official license fees and taxes. By the mid-1880s, as revealed by Suzuki's careful research, there was a subtle but clear governmental policy shift towards an unofficial recognition of private silk reeling firms. The industry began to recover and grew modestly that by the mid-1890s, there were a little over ten factories with about three to four thousand reeling basins.

The mid-1880s policy shift did not constitute a major ideological change in the Self-strengthening movement that discriminated the private sector. In cotton spinning, for example, the government continued to prohibit private involvement largely because the government itself had been planning a mechanising spinning plant. Similarly, attempts by Shanghai silk reelers and merchants in the same year to introduce steamships into inner rivers were also thwarted by local officials who were protecting the interests of traditional shippers.

The major ideological shift only came after China's defeat by Japan in 1896, which eventually led to the full lifting of legal constraints against private enterprise. In contrast to the difficulties of machine-reeling in Shanghai, the response of the traditional hand-reeling sector in the rural Lower Yangzi to the new foreign demand seemed to be far more successful.

To meet the European, particularly the American demand for evenness and uniformity, a new type of merchant firms, the so-called Jinsihang, centred in Huzhou, Zhejiang province, emerged in the 1870s to organise a re-reeling process where the hand-reeled silk collected from peasant household or silk hongs was first sorted by type, colour, and grade. Then on a putting-out basis, these sorted silks were redistributed to the peasant households for re-reeling using newly invented re-reeling tools. The Jinsihang then packed these rereeled silk for the export market.

Although still an imperfect substitute for machine-reeled silk, it did not require technical and institutional change beyond the traditional framework. The re-reeled silk was a clear improvement over the traditional silk in the overseas market and was important in explaining the resilience of handreeled silk exports from Shanghai in the 19th century.

A few years before John Major arrived in Shanghai to start his ill-fated filature, a poor and struggling part time school-teacher named Chen Qiyuan, left his native county of Nanhai in Guangdong province for Southeast Asia.

For the next ten years in Vietnam, Chen seems to have not only amassed a small fortune from his retail trade, but also discovered the French style filature operating in Vietnam. It was said that Chen travelled extensively in Vietnam and Thailand to carefully observe the machines, marking down their sizes and structure, often surreptitiously, using his body as a frame of reference. In 1871, Chen returned to his native town with an elaborate plan to start a modern filature.

But unlike Major, Chen knew China or at least his hometown. Beware of the kind of possible interference from the local government and guilds, he decided against such conspicuous location as the city of Guangzhou (or Canton) for his factory site. Instead, he set his factory in his familiar hometown, the Nanhai County.

To fit a modern mechanised filature into a rural setting, Chen had to solve a series of technical problems. He reproduced the French style reeling machine with the two most crucial innovations of the European filature, the double-twisting mechanism and steam-heated oven. But he largely replaced the iron structures with locally available materials such as wood, bamboo and china. He used the foot-powered treadle to replace steam power and, to add a Chinese touch, used all-purpose Chinese chopsticks to pick out cocoons from the boiling basins.

To construct the machine and especially to adapt and install the steam boiler in the village, Chen engaged an ingenious traditional artisan in Guangzhou, who later became a main supplier of reeling machines in the area. The modified machines not only can be locally supplied but also much cheaper, at only about one fifth of the value of a Shanghai style machine. Mobilising the lineage and village organisation, he successfully recruited female workers in the neighbouring areas. Using credit advance and other tools, Chen managed to procure cocoons from the neighbouring markets. To market his raw silk, Chen set up special trading firm to sell his raw silk directly to Western merchants.

Modern filatures in Guangdong benefited from the multivoltine silkworms reared there, which would hatch four to six times a year in contrast to only one or two times from the uni- or bi-voltine silkworms in Japan and the Lower Yangzi. Although somewhat inferior in quality, multivoltine silk production saved the finance cost of cocoon purchase and allowed the year-around operation of modern filatures. The factory was started in 1874 after almost a year of preparation and construction. Unlike the Ewo in Shanghai, Chen's factory was said to have turned out profits even in its first year of operation. The filature product, much more smooth and even than the traditional silk, was well received, fetching a price a third higher than the hand-reeled.

Within a couple of years, Chen's factory expanded to about 800 basins. More remarkably, Chen was keen to diffuse this technology, receiving, as he claimed, over a thousand visitors during the first three years of operation. As a result, modern reeling factories spread quickly in Nanhai and in the neighbouring Shunde County. By 1881, there were already about 11 factories in Nanhai and 6 factories in Shunde, with the total of reeling basins exceeding 4000.

Understandably, China's first factory encountered various resistances from both the vested interest and traditional-minded locals. But Chen, through provision of local charities and participation of public projects, managed to pacify the discontent and win the trust of the village elders. The growth of filatures occurred in Guangdong largely under China's characteristically ambiguous legal environment, which Chen Qiyuan took full advantage of using his social network. That ambiguity, however, came back to haunt the industry.

On October 5th of 1881, a sacred day of sacrifice for the Nanhai silk weaver guild, more than a thousand weavers marched to destroy and loot a filature in the Nanhai County. On their way to Chen Qiyuan's factory, they broke into violent clash with the local villagers defending the factory. The violence was said to have claimed two lives and wounded hundreds. Although the weavers had long harboured anger towards the filatures, their Luddite violence was triggered by a particularly bad year of business that they blamed on the filatures for taking away their source of cocoon supply.

Eventually, local county militiamen had to be mobilised to put down the riot. The magistrate, while punishing the weavers for the violence, also ordered all the filatures in the county to shut down immediately. The magistrate's ruling, backed also by the provincial leader, was in line with the ruling ideology of the Self-strengthening movement that discouraged private initiatives in modern industry. Appeals by Chen Qiyuan and others against the ruling failed. In 1882, Chen had to move his factory to the Portuguese colony of Macao. Interestingly, filatures in the nearby Shunde County survived under the protective wings of the local gentries there.

Fortunately, the mid-1880s policy shift of the Self-strengthening movement towards private modern reeling firms came in the right time for Guangdong. Around 1886, the Qing government began to levy taxes on the sale of machine-reeled silk, which is tantamount to an unofficial recognition of these enterprises. Three years after the forced departure out of Nanhai, Chen's factory was allowed back in Nanhai County. But by then the Shunde County was already on its way to become the center of growth. By 1894, Guangdong exported ten times the amount of filature raw silk in 1880.

At the time Chen Qiyuan designed the steam-heated reeling machine, he also made a model complete with the French Chambon double crossing system, but used charcoal stove instead of centralised steam boiler for heating. This model, which did not require concentration of equipment in a centralised location, was an instant success. It diffused quickly among individual peasant households and was later adopted by merchants in a variety of organisational forms that included putting-out and "manufactories" system. Part of these "improved" hand-reeled silk went for export but the bulk of them supplied the domestic market partly.

Japanese modern filature started exactly where the first Shanghai Ewo filature left off. Jardine trading firm were considering to set up a modern filature in Japan in the late 1860s. Meanwhile in 1869, the Maebashi han sent an official, Hayami Kenzou, to Yokohama to investigate the foreign raw silk market demand. Visiting the Swiss embassy, Hayami was shocked to find that Japanese raw silk was selling at half that of the price of the Italian and French raw silk in London. This brought to home the urgency of improving the quality of Japanese raw silk adopting European methods.

In 1869, the year Major died in the rural Lower Yangzi, Maebashi han officials visited the Yokohama office of the Jardine and soon after started the negotiation of transferring the reeling machines of the ill-fated Shanghai Ewo filature for starting up a joint-venture in Maebashi. The negotiation did not fall through. The han officials decided to independently start a modern filature with the purchase of Italian style machinery from a Swiss trading firm in Yokohama. Approved by the newly established foreign ministry of the Meiji government, the officials employed a Western technical supervisor recommended by the trading firm.

Hence, Japan's first modern filature started in 1870 with 12 filature machines. The machinery was initially set up in the middle of Maebashi in June. But it encountered stiff resistance from the local officials and merchants who feared for the loss of their handicraft interest. Three months later, the filature had to be relocated to the isolated village of Iwagami with Hayami himself having endured threats of assassinations in the process. The machine adopted the Italian self-crossing Tavelle system and centralised steam-heating, but was greatly modified to suit the local conditions - wood replaced metal for the reeling basin, human and later water-power rather than steam power was used.

During the four-months when the technical advisor, Casper Mueller, was under contract, Hayami and a few other people worked around the clock to grasp the technology. The Maebashi filature never truly became a viable business. With the abolishment of han officialdom under Meiji's centralisation policy in 1872, it was taken over by the Gunma prefecture and later sold to a merchant house, the Ono-gumi. But in 1871, the Ono-gumi already launched a sixty filature reeling factory in Tsukiji of urban Tokyo, partly because, Mueller, who was hired again as its technical supervisor, feared going to the rural countryside. This urban-based private operation was even more shortlived and closed down in June 1873 partly due to the disappearance of mulberry fields and cocoon supply in a rapidly urbanising Tokyo.

These two early acts of introducing European reeling technology were followed by the third and the most well-known effort by the Meiji government itself: the erection of Tomioka silk reeling mill in Gunma prefecture in Oct. 1872. This government-owned-and-operated factory, with 300 all iron-made reeling machines, driven by steam power and housed under a specially constructed Western style red-brick building, became an imposing structure set in rural 19th century Japan. It also brought in Paul Brunat as the French head advisor along with 11 French experts, viz. 2 foremen, 3 assistants, 4 female reeling supervisors, 1 machinist and 1 doctor.

Despite its government sanction, the construction of the factory met with local xenophobic resistance. When the factory started recruiting workers in March of 1872, few people actually applied. It was only through government's repeated assurances that the factory managed to employ female workers on a two-year rotation basis from various parts of Japan. The Tomioka's transfer of the full bundle of European technology, irrespective of local conditions, was reminiscent of John Major's Ewo filature backed by Jardine's financial capital. And like Ewo, the Tomioka factory was a financial failure and was sold-off in the 1880s by the government at a value below the original investment.

The trajectory of Japanese reeling technology and organisation was to follow the path pioneered by the Maebashi and the Tsukiji type of filatures rather than that of the Tomioka. In August 1872, even before the closure of its urban-based Tsukiji filature, the Ono-gumi launched a new filature in the rural upper Suwa area of Nagano prefecture with some technical supervisors transferred from the Tsukiji plant. The Ono-gumi also engaged in financing several other filatures in the rural Nagano. In 1874 when the Ono-gumi itself went bankrupt, family based filatures were already multiplying in Suwa, which was soon to become Japan's centre of silk reeling industry.

However, most of these family based filatures were fairly small in scale and often produced an amount of raw silk below the minimum order from the retail merchants based in Yokohama. To overcome this problem, these rural based filatures organised "Sha," a type of village or sometime lineage based cooperative, each usually consisting of between 10 and 20 individually owned and operated filatures. The members of the cooperative would supervise each other's production to ensure homogeneity and standards in raw silk quality and jointly market their finished raw silk to Yokohama bearing their "sha" label. They also coordinated the purchase of cocoons and employment of female labourers. Moreover, the cooperatives were units of what we now know as joint-liability lending, through which they acquired direct credit advance mostly for cocoon purchase from the silk retail merchants in Yokohama.

By the second half of the 1880s, most of these cooperatives were equipped with a joint re-reeling plant that specialised in sorting and re-reeling raw silk produced by the individual member filatures to achieve greater standardisation for shipment to Yokohama.

This system of production, combining indigenous social organisation with borrowed intermediary technology, supported the rapid growth in exports of machine-reeled silk in the 19th century. By the 1890s, there are signs that this system may have outgrown itself. In 1895, the filature group under the Katakura family withdrew from Kaimeisha, the largest cooperative in Suwa, and independently started its own reeling factory with 360 basins, marking the beginning of the rise of independent manufacturers. With the growth of filatures, it became increasingly difficult for cooperatives, with their disparate production units sometimes employing difference power source, to rely only the final process of re-reeling for standardisation. In the case of Katakura factory that had consistently produced relatively high-quality silk, there was strong incentive to strike out on its own to fully capture the quality premium.

Equally important to this Japanese transition to independent manufacturers was the surge of filatures in Shanghai following the lifting of legal constraints against private capital in modern business in 1896. These relatively large-scale, independent filatures in Shanghai were turning out machine-reeled raw silk with a good reputation in the international market. Hence, the rise of modern factory is a direct response to overseas competition.

It is important to recognise that the divisibility of the bundle of European technology could, as in the case of Guangdong, also exert transformational effect on Japan's hand-reeling industry. In the 1870s, Japanese reelers already began to improve their traditional hand-reeling machines through the addition of the foot-treadle, the cogwheel and the crossing system. Without the use of steam-heating and inanimate power source, this improved hand-reeling machine (referred as improved Zaguri) did not require operation in a centralised location and hence, diffused rapidly among individual peasant households.

To satisfy the Western demand for uniformity, traditional hand-reelers and merchants also adopted the re-reeling process. While organisational forms such as putting-out was adopted on a fairly substantial scale, the prevalent institution of production was the cooperative system similar to that of the machine-reeled sector. Ironically, it was in Gunma prefecture - the region where Tomioka was located - that rose to become the centre of cooperative-based hand-reeling production.

Each cooperative, like in the machine-reeling system, shared centralised joint-rereeling plants but the initial reeling process was done usually at the individual household rather than the filature level. They also coordinated in the hiring of workers, marketing and credit procurement. In comparison with machine-reeling production, their scale was smaller and their production more self-sufficient, with cocoons often raised domestically. In spite of the above, the improved hand-reeling production proved to be a viable and important system of production for the 19th century Japan.

Regional Differences

Innovative technological and institutional adaptations of European technology to the local physical and social environment ensured the rapid diffusion of machine-reeling in Guangdong and Japan in the late 19th century. This success, contrasted by the laggard diffusion of machine-reeling in Shanghai due to the absence of technological adaptation, reveals the important role of creative technical learning for the growth in silk exports. Shanghai, through foreign direct investment, and Japan, through the efforts of central and local governments and private entrepreneurs, have been exposed to the most advanced Italian silk reeling technology. In contrast, Guangdong, through a private individual like Chen Qiyuan, learnt of the French technology through a second-hand or a thirdhand source. There is a possibility the Western reeling technology in Vietnam witnessed by Chen may have been indirectly transferred from Western experiments in Bengal, India. The cogwheel mechanism, which could have been dropped in the Bengal adaptation, was conspicuously absent in Chen's drawings. More importantly, the French chambon system adopted was relatively outdated compared to the Italian Tavelle system, but Guangdong was to stick with it all the way into the 20th century. Apparently, the channels through which knowledge was transferred had long-term productivity implications.

However, machines were only part of the story. Machine-reeling has shifted the reeling process out of individual peasant household. In comparison with traditional hand-reeling, modern filatures are far more dependent on market transactions. Cocoons consisted of 60-80% in the valued-added of raw silk. As is true throughout China and Japan, modern filatures relied heavily on short-term finance. The growth of silk-reeling in Suwa, Nagano, for example, was a tale of financial expansion. In the 1870s and early 1880s, a financial network based on credit advance linked the Yokohama silk retail merchants with reeling cooperatives in the mountainous Nagano. This system of credit provision relied on the use of bills of•exchange and the network of national banks set up and diffused across rural Japan since the Meiji.

By the late-1870s, the issuance of insurance contracts by the domestic land-transportation corporation further reduced the risk of lending. In the mid-1880s, the Yokohama Species Bank and later the newly established Bank of Japan provided low-interest indirect finance to the silk reelers through rediscounting bills held by Yokohama merchants and local banks. General decline in interest rates would lend itself to higher level of machine-reeling production. But the interest rate story is also part of the transaction cost explanation posited in our model. An essential element supporting the expansion of Japanese reeling industry in the nineteenth century was the rapidly improving external infrastructure built up in the Meiji period.

Newly built modern communication system such as telegraph between Tokyo and silk-producing regions reduced the power of local silk merchants and enhanced the interests of silk-producing cooperatives.Infrastructure construction or improvements in roads, shipping and particularly railroads, which went on continuously throughout the 1870s and 1900s, not only served to link the rural silk producing regions with Yokohama, but also helped filatures to extend their cocoon purchase to locations farther away from their immediate surroundings.

In the Lower Yangzi, a dense financial network composed of foreign banks, trading firms and native banks clustered in the treaty port of Shanghai, but financial intermediaries beyond Shanghai were far and few until after the 1920s. At each cocoon purchase season, silk-reelers had to organise shipments of bulky silver ingots guarded by military boats to the rural Lower Yangzi for cocoon procurement. The building of railroads had to wait until the early 20th century in the Lower Yangzi. It was even later in Guangdong.

To fully understand the differential rate of decline in transaction costs, one cannot avoid the large political economy story. Both Guangdong and Lower Yangzi turned out to be hazardous and sometimes dangerous environment for capital investment in concentrated locations. Due to the insecurity of private investment in machine-reeling, location wise, Shanghai had become a centre of reeling industry for all the wrong reasons, enduring high transaction costs almost from the beginning to the end.

In the clash between traditional handicraft and factory interests, Qing's initial stance in favour of the traditional was not merely ideological. In the latter half of 19th century, public finance in Qing China deteriorated rapidly and government revenue grew dependent on commercial taxes such as the likin tax. Merchants suffered increasing squeeze. In return for tax revenue, merchant guilds, through tax-farming, extracted the government for monopoly power, a process where both parties ended up being captives of vested interest.

In this regard, the retreat of Japanese mercantile control in the 19th century was highly illuminating. In 1873, in order to arrest the trend of quality deterioration in raw silk export, the Meiji government promulgated the Regulation on Raw Silk Production, which stipulated that all raw silk bound for export would have to bear government printed labels and all silk merchants were required to obtain license issued by the Ministry of Finance. Subsequently, the government organised a national Aratamei Kaisha

with branches all over the major silk producing regions. It became clear later that the licensed silk merchants entrusted with raw silk quality inspection and distribution of government labels soon dominated the local levels of Aratamei Kaisha.The Aratamei Kaisha, designed for quality control and possibly also as a blocking device against the intrusion of Western trading firms into rural Japan, had however, inadvertently led to the rise of privileged merchant group who, backed by government power, was beginning to exercise monopsony control over silk trade in Japan. In 1876, Hayami Kenzo, then an official in the government industrial promotion bureau, voiced to the government about his concern of Aratamei Kaisha becoming a hindrance to the development of the still nascent machine-reeling filature in rural Japan.

In 1877, the government officially abolished the Regulation, but it took another four years for the entire system of Aratamei Kaisha to whittle away. During this period, the government ministry stood firm against repeated attempts by local silk merchants to regain their trading privileges and their attempts to include the machine-reeled raw silk under their monopsony control. It was in the weakening of local merchant monopoly that paved the way for silk-reeling cooperatives and later factories to start direct trading relationships with the retail merchants in Yokohama.

The study of the evolution of the silk industries offered us important insights not only into the impact of political change on economic growth but also shed new light on the nature of traditional East Asian society. It is important to recognise that the transaction costs modelled in this chapter largely existed beyond the control of individual firms or households. Thus, to overcome them required coordination on a local and, in many cases, national level.

3

Silk Reeling Industry

Silk is a highly valued animal fibre used almost entirely for the production of high quality textiles. Sericulture i.e., the production of silk is a system of animal husbandry whereby the silkworms (i.e caterpillars from moths) feed on leaves to produce cocoons from which silk is obtained. The early stages of silk production offer particular advantages to small farmers in developing countries. It is mainly a rural, labourintensive industry requiring relatively low investment and offering high profit potential and foreign exchange. Production trends in major silk producing countries viz.. China, India, Japan, Korea etc., are given in Table 1.

Table 1. World Raw Silk Production (1998)

Country	*Quantity (mt)*
China	49430
India	14260
Japan	1080
Brazil	1821
S. Korea	146
Other	4990
Total	71727

Though silk accounts for less than 1% by volume of world fibre production, its relatively high value and high labour requirements make it an important industry in many of the countries mentioned above. Silk production and trade in most major producing countries are subject to close government control. Sericulture and production of silk goods are a complex agroindustrial process needing considerable agricultural, processing and organisational skills.

Sericulture in India

Sericulture plays a unique role in Indian economy. Its uniqueness rests in the fact that sericulture activities not only engage rural households in the cultivation of mulberry and silkworm rearing but also encompass in their fold a whole range of reelers and weavers. On every count sericulture needs to be expanded in India. Its main advantages are

(1) High employment intensity of 13 man-years per hectare per annum,

(2) Drought resistance nature of mulberry,

(3) Regularity of income to farmers and reelers, many times during the year,

(4) Its suitability for small farmers,

(5) Gainful employment to women folk in the household,

(6) Being a luxury final product, the transfer of income takes place from the rich to the poor and

(7) High export possibility creating trade surplus.

India is the second largest producer of raw silk in the world next to China. India produces all the four varieties of raw silk viz., Mulberry, Tasar, Eri and Muga. Tasar, eri and muga are called non-mulberry silks. Mulberry constitutes 90% of the total silk production of India.

In India, silk production grew rapidly between 1960 and 1990. Growth has been slow since 1992. Mulberry silk is chiefly produced in 5 states viz. Karnataka, Andhra Pradesh (A.P), Tarnilnadu (T.N), West Bengal and Jammu & Kashmir, contributing to about 99% of the total mulberry silk produced in the country. Interestingly, the states of A.P and T.N with almost no silk production during 1960, currently occupy the second and fourth position respectively. West Bengal at present contributes about 8% of the total silk production, while Karnataka contributes the lion's share (62.7%) with Andhra Pradesh and Tamilnadu contributing 23.4% and 4 3% respectively. (Table 2).

Table 2. Statewise Raw Silk Production in India (1998-99)

State	Production in mt	%of contribution
Karnataka	8944	62.7
Andhra Pradesh	3335	23.4
Tamil Nadu	616	4.3
West Bengal	1118	7.8
Jammun & Kashmir	86	0.6
Other	161	1.2
Total	14260	

Prior to independence, sericulture in India was always considered as a subsidiary occupation of poor farmers while rich farmers used to do only the cultivation of grains, pulses and orchards. The lack of technological know-how and low productivity of indigenous races were the main reasons for poor returns. In 1947, the silk production was limited to 800 tonnes.

During post-independence, attempts were made to transplant the temperate sericulture technology in the silk producing states like West Bengal and Karnataka, but it failed to bring any positive result. Having realised the need for their own technology, the Central Silk Board (CSB) established a number of sericulture research institutions in 1960s. With systematic efforts, it became possible in 1970s to develop a technology suitable for tropics.

New mulberry varieties coupled with agronomical practices were made available to the farmers. Package of practices were developed for silkworm rearing, besides realising new bivoltine races. Popularisation of the bivoltine races by the central and state governments of southern states brought a quantum jump in productivity. Since seed preparers started using bivoltine as a male parent for the preparation of multivoltine hybrids, the traditional poor yield crosses have been replaced to the extent of 85%. Consequent to this, mulberry sericulture was spread to non~ traditional states like Kerala, Maharashtra, Rajasthan and Gujarat in the 1980s.

While other crops (grains) wither away due to very little precipitation, mulberry survives such acute situation where ground water is also not available for raising the crops, thus providing subsistence to a large number of farmers. In fact, India has developed sericulture as an effective tool for economic development of the rural areas where irrigation facilities are available as well as for drought-prone areas. Sericulture has been introduced in the non-traditional states and also the states, which were once thought to be unsuitable for silkworm rearing, as a result of the special interest taken by the Government of India.

Sericulture in Karnataka

The beginnings of sericulture in Karnataka can be traced back to the last quarter of the 18th century. Tippu Sultan, the ruler of erstwhile Mysore state was instrumental in introducing sericulture in the state, since he needed high-value low volume commodities to trade with Europe in exchange of aims and ammunition. His other venture was pearl-fishing. However, sericulture took roots in Karnataka due to the meticulous care he took in its propagation and issuing instructions even from the battle field. The rulers succeeding him in Mysore also continued the royal patronage to this activity.

Though Kashmir and Bengal were leading states in India in sericulture in the past, Karnataka has become the leading producer in silk, accounting for a large share of the mulberry silk production in the country. This was possible because of the mild climate, government patronage and the skill of Karnataka farmers. With due cognisance of the importance of sericulture industry in the economy of the state, the Government of Karnataka undertook an ambitious sericulture development plan known as Karnataka Sericulture Project (KSP) with a total outlay of Rs 80 crore which included a loan of US $ 54 million from World Bank for a period of 5 years beginning from 1980-81. The programme was launched with a three-fold objective:

(i) a substantial increase in silk production from 2300 mt in 1978 to 4500 mt by March 1985,

(ii) promotion of bivoltine technology to improve the quality and export prospects of Indian silk products and

(iii) extension of sericulture activities to new areas in the state and to new strata of rural poor.

In order to achieve this, the project envisaged plantation of mulberry in new areas under irrigated conditions, establishment of grainages, chawki rearing centres, cottage basins, modernisation of silk filatures, spun silk mills, provision of technical services through Model

Chawki Rearing Centres, Technical Service Centres, Training Centres, expansion of the arrangements for marketing of cocoons, strengthening research capability for evolving new races of silk worms and strains of mulberry etc. Among the infrastructural facilities provided. On- Farm Development Programmes (OFD) is one with a provision of Rs 6.64 crore.

The programme was a good instance of development strategy seeking to combine the often conflicting goals of growth of production, development of backward areas and improving the employment and income of rural poor. However, in terms of aggregate indicators like actual expenditure incurred by the state government and the physical targets achieved, the programme had been sluggish in its progress.

The growth of sericulture activities in these five years in Karnataka is of very great significance. Traditionally sericulture activities were restricted to a few taluks of south Karnataka. Under the KSP, it was decided to extend sericulture activities to non-traditional areas of the state in order to reap the benefits of these activities in terms of increased income and employment generation. The emphasis of the KSP has been on the development and extension of bivoltine silk programme, which has good export potential and in addition, on extending the programme of cross breed cocoons for meeting domestic demand.

Eleven out of the 13 districts where OFD programme for the establishment of cottage basins was under implementation, have achieved less than 50% progress. The implementation of KSP was an admixture of successes and failures. Sericulture is not just restricted to the agricultural activities of mulberry cultivation, silk worm rearing and seed production, but opens up into the sectors of silk reeling i.e.. rawsilk production, twisting, weaving, dyeing and printing industries.

They support the ancillary enterprises of reeling machinery manufacture and by-product utilisation comprising of spun silk yarn manufacture and pupae oil extraction. This provides employment for the skilled and unskilled labour force and nurtures entrepreneurship as well right across the country. The progressive movement along the chain of activities is not limited to conversion of form but value addition as well. The chain of activities involved in sericulture and silk industry is typically represented in the form of flow chart, which is given in Figure 1.

Trends in the Reeling Sector

Reeling fomis a vital link in converting the agricultural produce viz. cocoons into an industrial product - the fabric. Reeling involves a series of intricate processes while converting the cocoons into raw silk.

The reeling sector in India is cottage based and highly decentralised. There are different reeling devices in India viz., Charka Filature consisting of cottage basin and domestic basin and Multiend reeling basin. Charka reeling device is a crude contrivance for producing silk, which has minimum mechanisation and practically no gadgets. These charkas could be hand operated or motor operated. Of late, more and more charkas are being made motor operated due to increasing labour cost. The charka silk has distinct identity in the market and it is generally associated with poor quality. However, a small proportion of good quality charka silk is also a reality. Charka silk is largely preferred by the handloorn weavers.

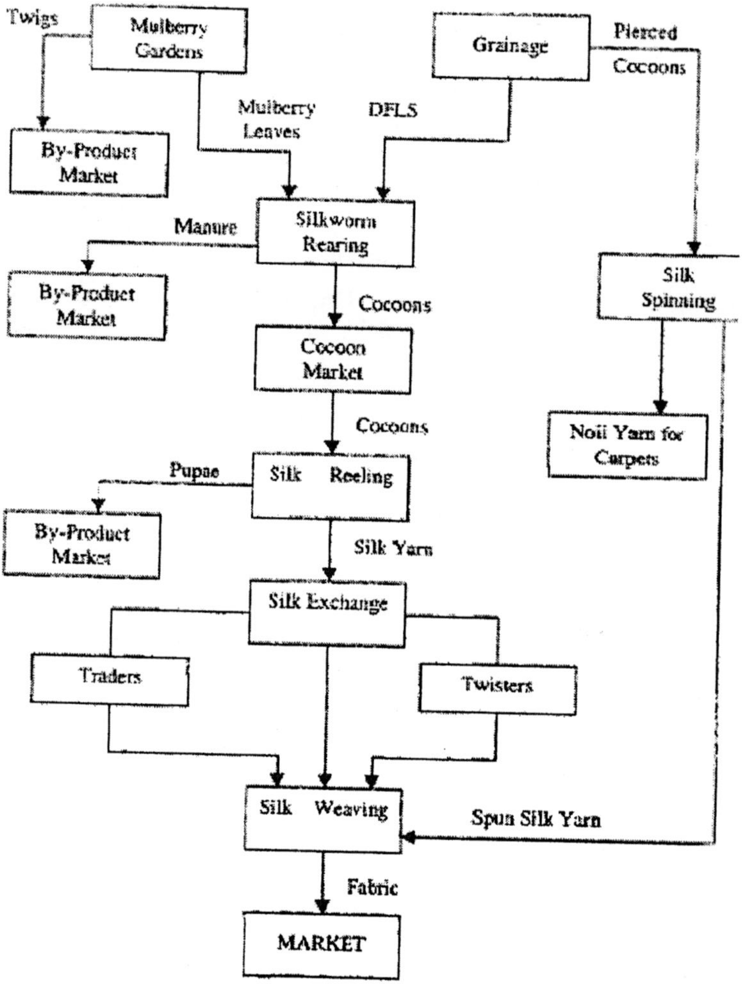

Figure 1. Flowchart of silk industry activities

In the intermediary level we have the domestic basin and cottage basin devices. These two devices in principle are the same. However, there is a basic distinction that the domestic basin silk is directly reeled on to standard sized reels, while the cottage basin silk is first reeled on to small reels and later converted into standard sized hanks entailing an additional process of re-reeling. However, the trade treats both the silks analogous and refers to it as filature silk. These two devices have comparatively better mechanisms and basic gadgets capable of producing better quality raw silk. The filature silk with wide ranging quality is produced in reality. They fetch higher prices than the average price of charka silk and are purchased by handloom weavers and powerloom weavers and the best quality by the zari manufacturers. The capacity to purchase cocoons in the case of filature reelers is better than that of charka reelers.

The upper end of the reeling technology has automatic and semi-automatic reeling machines and multiend reeling machine. The automatic and semi-automatic reeling machines have been a failure in meeting the economic, productivity and quality requirements of the industry owing to its huge investment cost and serious limitations of the basic raw material viz. cocoons.

Multiend reeling machine is an improvement over the cottage basin and is the most appropriate technology that can handle the available quality of cocoons to produce gradable quality raw silk. Multiend reeling machine has been a new phenomenon in the Indian silk reeling industry. However, the investment cost is very much on the higher side in comparison to cottage basin or domestic basin of equivalent capacity.

Silk Weaving

There are about 182,325 handlooms and 29,340 powerlooms engaged in the production of silk fabrics in India. Handlooms consume about 70% of the raw silk, while powerlooms consume the remaining 30%. Indian silk fabrics are predominantly saree material and are exquisite. Silk weaving is concentrated in a few places in the country, like Kancheepurarn, Kumbakonam and Ami in Tamilnadu, Mysore, Bangalore and MoLkalmur in Karnataka, Dharmavaram, Pochampalli and Prodattur in Andhra Pradesh, Varanasi in Uttar Pradesh and Murshidabad and Birbhum in West Bengal.

Each of these regions produces distinct designs and textures and is mainly dependent on raw silk from Kamataka. In contrast to most silk producing countries, India produces an infinite variety of small fabric lengths compared to the mass production scales elsewhere. In India, nearly 75 % of the silk is yam dyed and 25 9c piece dyed. The strength of Indian silk lies in its decentralised nature and its ability to produce limitless varieties of fabrics in small quantities with numerous designs. These attractions have helped to expand silk consumption amongst wider income groups. Urban population absorbs the various sarees and silk fabrics.

Market Structure

Cocoon and raw silk markets are two decisive segments of the silk industry. In the cocoon market, the farmers are the cocoon producers whose wares are bought by the reelers. The reelers who are buyers in the cocoon market are sellers of raw silk in the raw silk market to traders, twisters and weavers. The cocoon and raw silk markets have certain unique features that distinguish them from the traditionally known market structures.

The size and number of buyers and sellers, the ease of entry and exit, the demand and supply gap and its impact on prices etc, present a distinct scenario in Indian silk industry. How far these features differ from a traditional economic market structure and to what extent do they resemble it, would help the understanding of the special features of Indian silk industry in a proper perspective.

Silk reeling industry does not fall in the category of either monopoly or oligopolistic competition. This is because the silk reeling industry is characterised by the presence of a large number of buyers and sellers, both in cocoon and raw silk markets. Though it has similarities to that of monopolistic competition in the form of product differentiation and price differences, these are not the results of deliberate attempts on the part of producers and there is neither advertising nor any sales promotion activity undertaken by a seller.

Even the price differences occur due to the differences in quality as perceived by the buyers. Another distinguishing feature is that the market for cocoons as well as raw silk are facilitated by the government, which implies market intervention that is absent in traditional market structure of monopolistic competition.

However, the silk industry has quite a few characteristics, which make it comparable to that of pure competition. Therefore, it would be appropriate to assess how far silk industry is similar or dissimilar from that of pure competition. In this regard, a broader description of the functioning of cocoon and raw silk markets is necessary. The cocoon and raw silk markets of silk industry appear to be two competitive markets working in tandem. The two are interdependent to such an extent that they cannot be looked at in isolation, since what happens in one market is likely to influence the other.

There are 56 government cocoon markets in the state. At least 10 markets have been identified as major (grade I) markets, while the remaining are grade II markets. Here, farmers bring the cocoon lots and display them one beside the other in trays provided for the purpose. The availability of trays is on 'first come first served* basis. The government bidder starts the first bid i.e. the government bid and later on the reelers (buyers) pick up from there.

The highest bidder would be given the cocoons, provided the rate is agreeable to the farmer. The buyer is required to be a registered reeler or a registered agent of the reeler. He is also required to keep a safe deposit with the authorities, to enable his participation in the auction on any given day. It is estimated that around 80% of the cocoons produced in the state is transacted in the cocoon market.

Similarly there are nine silk exchanges in the state and Bangalore exchange is the major market accounting for nearly 85% of the raw silk transacted in silk exchanges. The price signals from this market are caught up almost through out the country and forms the basis of their respective market prices. Here reelers bring raw silk in sacks that are poured out one after the other and the price stabilising agency starts the bid price. The agency is supposed to be working in the interest of the reeler. The twisters, traders and weavers participate in the auction as buyers.

Unlike the cocoon market, money does not change hands between buyers and sellers in the silk exchange, though the policy remains the same. The relevant market fee is paid to the government and the reeler would have to go to the doorsteps of the buyer to get his amount. It is assessed that hardly 20% of the raw silk produced in the state is transacted in the silk exchange. In spite of this, it appears that, private transactions are led by the prevailing rates at the silk exchange.

Sellers in the Cocoon Market

There are 350 thousand farmers involved in cocoon production i.e. sericulture. This reflects the large number of sellers. Land holding of the farmers are quite low and also mulberry cultivation has to compete with other food crops for its share of acreage. The largest land holding reserved for mulberry of an individual farmer is around one acre and the output per annum would be 500 kg of cocoon, which forms such a small percentage of the total supply. Even if this farmer envisages a ten fold increase in cocoon production, his contribution towards the total supply continues to be a very small percentage. This means that even big farmers individually cannot influence the market situation in terms of quantity or price.

Sellers in the Raw Silk Market

There are around 24 thousand reelers (as per the registered number of reeling units upto March 97) in the state. They all own small reeling units and the maximum size of a reeling unit does not exceed 20 basins. The cocoon requirement for a period of one year by this unit is 60 mt, which is less than 0.075% of the total supply of cocoons. Even if his capacity to consume cocoons increases tenfold, his requirement would shoot up to 600 mt, which still forms 0.75% of the total cocoon supply. Thus the individual purchaser would not in any way influence the market situation. These reelers who are buyers in the cocoon market are sellers in silk market.

The quantity per seller is quite small since the reeling unit sizes are small and working capital requirements being very high, reelers are desperate to encash their produce early leading to smaller cycle time. Even if we envisage a tenfold increase in the output of a particular reeling unit, it hardly becomes 0.7% of the total supply of raw silk and this quantity cannot influence the market price. They can only adjust to the market price and hence are price takers. This large number of sellers and buyers ensures that individually they do not cause any perceptible impact on the price. Neither the producer nor the buyer has much say on the price and the market conditions determine the cocoon price.

Buyers at the Silk Exchange

There is an element of peculiarity as far as the number of buyers at the silk exchange are concerned. The number of buyers here gets reduced as compared to the number of farmers or reelers, owing to the presence of traders. Individual twisters and weavers in turn depend on traders. It is estimated that there are around 600 buyers in Bangalore silk exchange, while there are 12,000 registered sellers (reelers).

However, hardly 20.7c of the total quantity of raw silk, get transacted in the silk exchange. Since silk is a high value item, no single buyer would be able to make a purchase that is significant in comparison to the total supply in the market. Also, majority of the silk consumption is by handlooms and the powerlooms, which axe very conventional, operating at low speeds. The productivity in both the cases is thus quite low and the purchases per unit will therefore not be in large quantities since the weaver would not want a huge inventory or work-in-Process.

The reasons that might have led to the feature of existence of very large number of small players are perhaps the inherent drawbacks of the silk industry. The scale and scope economies are negligible and this could be the reason for the existence of many small firms. This perhaps has eliminated the motivation for concentration and increase in the size of firms.

Ease of Entry into and Exit from the Industry

As far as the farmers are concerned, investment in sericulture is minimal and they always have the choice of switching over to any other food crop from sericulture. But such decisions cannot be on a short term basis, since mulberry is a perennial crop with a life span of at least 10 years. However, they may choose to downsize their activity if the situation warrants. In the long run, uprooting mulberry for other food crops is quite common in certain areas, when the market situation is not conducive.

The equipment of the reeling and weaving sectors in the silk industry is notably simpler and less capital intensive than that of other manufacturing industries. The sunk costs are thus minimum. Contractual obligations are also minimal and vertical integration or one to one interaction between buyers and sellers is a far cry. These aspects make the entry into and exit from the industry very much easy. There are no barriers, implicit or explicit in terms of legal, technological, financial or others, for anyone to start an enterprise in the silk industry.

The exit/entry phenomenon is quite transient to a majority of the units while very few of them would think in terms of total withdrawal. Also, since the fixed costs are minimal, the producer would emphasise on the variable costs and would not like to go ahead with production when he is not able to recover his costs from the sales of silk and by-products.

Such decisions of stalling the production would not in any way hamper his continuation of production in the future or his reputation. It so happens many a time that reelers pay minimum wages to their labourers during periods of closure, since they are better off then, rather than engage in production.

Product Differentiation

The product either the cocoon or the raw silk is virtually a standardised product. Pairing of buyer and seller is generally by chance and not by preference. The buyers look upon the products by various producers as perfect substitutes. The deviation here is that though at a macro level, the product appears as a homogenous commodity, the situation is however complicated by the dimension of quality.

The quality of cocoon can be perceived as a function of its size, colour, compactness, freedom from defects and such other attributes, which in turn can be linked to the disease free layings used, rearing conditions and practices, etc. The raw silk quality can be grasped by its colour* uniformity of size, lustre and feel. The source in both the cases also becomes an indicator to the buyer about its expected quality. The most that can be said to offset this complication is that the range of quality is open to inspection and that commodity could be said to vary in a known fashion. Grading before transaction and uniformity of price for a given quality is also not streamlined.

Another deviation is that certain amount of pairing of buyer and seller would take place by preference associated with the range of quality offered. Prior experience of buying from a particular producer or an area would come in handy. Though it is considered that non price competition is absent in a pure competition model, it is very much a reality in the market situation where the lots of cocoons are displayed side by side and comparison is inevitable and perceived quality is directly reflected in price varying from lot to lot. Attention to the quality of the merchandise is unavoidable. Though buyers in the cocoon market cannot spell out their requirements overtly, they have the freedom to choose from among the lots, in terms of what suits them the best.

Perfect Knowledge

The buyers and sellers have access to information on price changes. All buyers are able to see what prices are being paid for what quality and the way they varied. The order of displaying the cocoon lots for auction is on first come first served basis, so that the distribution of large and small lots and of good quality and poor quality lots would be random over the period of sale. Since the entire transaction is open for anyone to sense the market trend, information on the prices at which the cocoon lots are bought, the quantity transacted or the quantity offered is accessible by anyone interested.

The farmers, reelers and weavers always keep track of the sales of eggs, the rain, season and demand in the fabric market and such other relevant movements along the cross section of the market. There are, however, no means for either subjecting such information to quantitative analysis or making them cumulative so that forecasting is possible by a larger section of people, so as to help them in decision making. It is up to the individual players to make a quick qualitative analysis of the situation. Also, the system does not provide for a mechanism to receive feedback on quality or for expressing the buyer's requirement at the time of purchase either for the present or for a future transaction.

While the demand or supply facing the individual competitive firm is perfectly elastic, it is quite inelastic as far as the total supply of the industry is concerned. Irrespective of the price of the cocoon, the entire quantity of cocoon that comes to the market is sold and on record there has never been an instance in the history of the cocoon market that cocoons have been left unsold.

Demand may therefore be viewed as potential demand and activated demand. At any given point of time activated demand equals the supply and at the industry level the demand is less elastic. However the number of reelers/buyers gets reduced. The demand at the level of the individual reeler gets retracted temporarily when the supply is less and, in turn, is more elastic. The quantity of cocoons demanded by the individual reeler depends on the plant capacity, labour availability, working capital availability and revenue from silk sales.

Reelers have to buy in a competitive market and also sell in a competitive market and this places them in a sensitive position, while the farmers are fairly robust, in the given situation. Aggregate supply or availability of cocoons to the industry as a whole in the long run may remain relatively constant, but its distribution at the point of requirement, spatially or temporally may get redistributed.

Demand for Silk

The production of silk textiles forms a negligible portion of the total textiles production in the world as well as in India. The natural fibres constitute 47.4 % of the total world textiles production and silk production is about 0.42% of natural fibres and 0.2% of the total textile production.

Silk thus forms a miniscule in the total textile sector in the Indian context as well as world over and textiles as we know is one of the three basic needs of humankind. Silk has been an ancient fibre and has many reasons to be the most ecologically favoured textiles. Silk fabrics are not purchased as a necessity of consumers but more as a luxury, which makes it more a niche market than a common man's market. Silk is valued for its aesthetics and comfort properties. Price is one of the major determinants of demand for silk and rise in income levels and living styles over time have caused an increase in demand for silk.

There have been attempts for generic promotion of silk the world over. In view of its unique features and miniscule production, product substitutability ceteris paribus, is no threat- Even with reference to the aspect of change in consumer preference, the change boils down to differences in the value addition process in terms of fabric construction, finish, colour combinations and design content and this does not generally call for major production shifts or product differences.

Capacity of the consumer market to absorb silk either in the export market or the domestic market is not a threat, provided better quality silk is made available at lower price. A downward price trend would bring in more and more consumers under the silk fold. The fact that large quantities of Chinese silk are being used for fabric manufacture in the Indian industry reinforces the fact that demand and hence the gap exists. It may be recalled that Chinese silk is of better quality and available at lower price.

The starting premise here, therefore, is that the demand for silk exists unquestionably at the consumer level. If this demand of the consumers is to be met, fabric manufacturers would naturally transfer this demand on to raw silk. Since raw silk is not a perishable item, fluctuations in quantity demanded due to seasonality can be taken care of, by adequate stocking in the form of raw silk.

Raw silk is the input for fabric manufacturing, irrespective of the changes in consumer preference and tastes. This implies that fluctuations in fabric demand need not induce changes in raw silk production pattern. This demand for raw silk is to be met by the reelers and who, in turn, have an invariable input viz., cocoons. There is no control over the yield or productivity of these cocoons in the short run so as to make it respond to the changes in raw silk demand. Raw silk production thus becomes more a function of cocoon production than fabric demand.

Demand for silk is therefore for the fabric manufacturer to perceive and enlarge, while demand for sustenance exists. The quantity demanded is an open-ended figure and it is up to the seller's capability to penetrate and expand the market by offering a wide variety of products which will not be radically different as in other market economies. If more weavers get into this activity there will be more demand for raw silk.

In a competitive market, left to itself without government intervention, raw material availability and price will dictate the number of weavers. If the number of weavers increases, raw silk supply will also have to follow suite. With increased demand for raw silk, more reelers will corne into production arena as much as the cocoon availability might permit. An increased demand for raw silk will lead more farmers into sericulture. Silk, as a product has no competitors. It may only have silk-like substitutes and no real substitutes.

The demand for silk fabrics thus may be considered very high and there exists scope for more fabric manufacturers to enter the industry, which can only be restricted by the supply of raw silk. The same holds good for the entry of reelers, where the supply of cocoons becomes the constraint. The number of farmers or area under sericulture would be a decisive factor for cocoon supply. The choice of the farmer depends on the revenue assured by sericulture. The industry thus heavily rests on the supply side.

Conventional wisdom attributed the growth of exports almost exclusively to the demand side, while this was not so in the case of silk. According to Federico the total growth is evenly shared between demand and supply side. He says that it appears difficult to assume that silk exports alone could have caused development and probably development led to the exports.

One would expect that in the absence of government intervention, the market conditions of supply, demand and price would decide the equilibrium and the whole system gets adjusted at the optimum level. However, the aspects of seasonally and wrong judgements may make equilibrium conditions difficult to attain and even when attained in the short run, make it quite short lived.

As one moves up the technology of production, the ease of entry and exit gets restricted. Ease of entry especially, at the low to middle level of technology coupled with the demand has led to mushrooming of reeling units. Then the question arises as to why the rearing sector has not been able to meet the requirements of the reelers, despite the governmental support in terms of floor price and other supportive schemes.

Floor price should result in surplus but shortage continues to persist. The fluctuations in price, seasonality factor influencing the cocoon yield and quality, the added component of risk due to crop failure or market indifferences as also the dependence on food crops for sustenance are some of the vital factors that seem to have restricted the matching expansion of mulberry cultivation. Tradition, risk averseness, environmental factors coupled with extension efforts become major factors influencing the decision of more farmers getting into sericulture apart from the price factor. This has helped them to remain in a more commanding position than the reelers.

Short run supply is relatively inelastic and the long run supply is much more elastic as the practical feasibility of expanding and contracting supply even in the long run is questionable in the rearing activity. Mulberry is a perennial crop and has a time lag to encash upon it. While in the case of reelers effecting changes in supply either in the short run or in the long run is very much easier.

Lack of vertical integration results in buyer requirements not being properly understood and therefore not met by the seller or the producer, because buyer and seller pairing does not take place in the organised markets by design but as a random occurrence.

With the introduction of cocoon markets, oligopsony was eliminated i.e. the buyer clout- the power of buyer to set the price. Intermediaries gaining extra profits at producer's expense an indicator of market imperfection was also eliminated, in the cocoon markets.

Deviations from Perfect Competition

Going by the number of registered reeling units up to March 97, the cocoon requirement per year of 300 working days, works out to be 128305 mt. The cocoon production for the year 1997-98 is 80,656 rnt and there is a scarcity of 47650 mt and a part of which may be met by the production outside the state. However, there still exists a huge gap. The supply constraint would restrict the reeling capacity utilisation to around 63%.

In pure competition, firms and consumers maximise either profit or utility on the basis of prices set jointly by the forces of supply and demand in decentralised markets. Sellers and buyers behave as if they are guided by an invisible hand. Their self-interest drives them to gather sufficient information, to make good production decisions most of the time and to correct their mistakes in the future when mistakes are made. Each player here expects to raise one's standard with each decision to buy, or produce or sell.

Competitive firms operate in an environment of anonymous rivalry. This anonymity forces competitive firms and consumers to depend almost exclusively on market signals for their information. Reeling units may shut down (cease operations altogether) even in the short run if the market price falls low or cocoon price become very high that loss from operating is greater than the fixed costs. When economic loss is greater than fixed cost, the firm is better off by not producing at all and accepting the loss of its fixed costs each period. The fixed costs mainly consist of labour cost since skilled labour cannot be dispensed with in the short run closures and investment on plant and equipment is so low that it hardly matters in majority of the cases.

Reelers do not enjoy economic profits to lead to the establishment of more and more number of reeling units. Owing to the fact that nearly 90% of the cost of production is that of cocoons, very little value addition takes place. Major focus of government support has been largely on the cocoon producers with little support to the reeling sector and the government bid ensures that the cocoon prices do not go below a certain level. With these features of reeling, what would then explain the presence of so many reeling units?

The establishment of a large number of reeling units has caused a big gap between demand and supply of cocoons. It appears that, the ease of entry to and exit from the industry, the sustenance in terms of the scope for continuous functioning and minimum returns the industry offers, access to raw material (though not in sufficient quantity) and ease of selling raw silk (though not at the market price), leads to the entry of more and more reeling units. Much of the market studies especially in the competitive market, emphasises the issue from the point of view of the seller (farmer) and the supply control required, so that, prices do not come down. But the situation in silk industry is different. Irrespective of the supply position, farmers tend to make profits. Much of the impact of price fluctuation is borne by the buyer i.e. the reeler. The cost of production of raw silk varies greatly between reeling units.

All the firms may or may not achieve production efficiency. Similar is the case with farmers. There are a lot of variations in both space and time. For example, access to markets, the intrinsic quality of the land itself, ease of cultivation, availability of skilled labour, water quality, cost of raw material, etc. would affect the yield, quality of the product and more importantly the input costs. Also some producers might have pursued innovation more than others and the timing of investments might have varied so that while some firms are using new modem capital equipment, others are using machinery and plant, which is old and obsolescent.

The excessive volatility of prices is often blamed as a cause for making reeling so risky. Reeling has been defined as "a speculation", whose results depend mainly on the price variations, often large, sudden and unpredictable. The changes that took place earlier may have repercussions not only for now, but also in the future. Consequently, decision making is based on a set of expectations and nobody can be certain whether they will prove correct. This is likely to lead to speculation. Expectations, speculations and errors - all features of uncertainty are then likely to influence the development of economic processes.

Though it can be safely termed that no individual buyer/seller is able to influence the market, the overwhelming role of the government makes it different. The government does not have control on the market conditions or prices as in China, but meddles with the market situation, not being sure of its impact. The government bid in the cocoon market, which almost represents floor price, the cocoon purchases by the government silk filature in large quantities by paying higher prices, the presence of Karnataka Silk Marketing Board (KSMB) in the silk exchange, which makes an attempt to have significant impact by purchasing a certain portion of market arrivals do lead to far from equilibrium conditions. Though the quantity they purchase is small in comparison to the total supply, they are significantly larger than any other individual purchase.

Though a perfectly competitive market system would lead to the attainment of economic efficiency, the conditions that have to be met if the economy has to be described as 'perfectly competitive' are extremely strict. They are not met, by a very long way in most of our economies. Firms operate in partial ignorance and under uncertainty rather than certainty, i.e., they do not know exactly their existing or future cost and demand curves.

In reality, the producers will not know the precise shape of the demand curves that face them in their efforts to sell, nor for that matter will they be in a position to anticipate correctly the magnitude of a shift in the demand curve. One area in particular where significant errors could occur would be if producers take a temporary increase in demand for a permanent one. The repercussions of such errors in anticipation would be amplified if the adjustment of production to demand also lead to increased demand for machines/capital. The effect of miscalculating or just misjudging the nature and size of demand variations may not only delay the establishment of a new equilibrium situation but it may also make the establishment of a new equilibrium situation impossible.

4

Silk Reeling Operations

Sericulture, as an agroindustry occupies an important position in Indian economy owing to its employment generation, production and exports. Its structure comprising cocoon cultivation, silk reeling and weaving of silk fabrics has strong intra-iinkages, having relatively high employment intensity and export orientation. Therefore, it would be an ideal sector to be used for India's rural development strategy.

However, sericulture industry has some typical problems associated with cash crops as also cottage industries in general. It appears that these associated problems from the two ends have a multiplicative effect leading to a state of instability in this agroindustry. But at the centre of the problematic issues is the task of understanding the problems of the reeling units. Poor quality of cocoons, very thin profit margin which easily gets whiffed off in adverse conditions persisting over a larger part of the year, fluctuations in cocoon and silk prices and absence of sufficient financial support, are some of the major problems encountered by the reeling units more or less, continuously.

Reeling sector is the backbone of sericulture industry, but has to confront the problems of poor quality of cocoons associated with a high price on the input side and competition from the imported Chinese raw silk on the output side.

Reeling sector is input oriented and its operations are influenced heavily by three factors - cocoon quality, cocoon supply and cocoon price. These factors along with the financial condition of the reeler influence the cocoon purchase decision. Seasonally in the cocoon market situation induces seasonality in the operations of a reeling unit, which tend to limit the earnings of a reeling unit.

There is hardly any institutional mechanism for fixing the price of cocoons based on quality and there are no organised procedures for cocoon testing and price fixation. There have not been adequate efforts to understand the micro problems of reelers from the points of view of quality, availability of cocoons, operations and the market for raw silk.

The developmental efforts in the Indian silk industry have often been of a repetitive nature since there has not been an integrated approach of documenting and analysing the problems in understanding its cyclic nature. Thus, there has been an absence of any effective mechanism by which experience can be made cumulative. One would look for a common approach to the repetitive problem rather than a repetitive approach to the common problem looked at in isolation.

Reeling is an important sector responsible for the conversion of cocoons into raw silk and in fact it is the weakest link owing to the fact that it is plagued with innumerable problems and the value addition that takes place is very low despite the vital transformation it provides. Reelers buy the cocoons at the cocoon markets. They use their personal judgement in assessing the quality of cocoons and accordingly, bid the price. The farmer has the choice to accept or reject the offer and participate in the next auction in anticipation of a better price.

In Japan and China, the situation is however different. The cocoons are tested prior to transaction and the cocoon price is fixed based on its quality and the prevailing raw silk price. Lack of a procedure for cocoon price fixation in India and absence of a standard specification for cocoon quality under Indian conditions are major drawbacks for quality improvement initiatives. The reelers face a fierce competition while purchasing cocoons, as there are a large number of contenders for the available quantity of cocoons.

The cocoon prices in India always tend to be on the higher side. The purchases made by the government filatures also have an influence over the cocoon price. The presence of government filature representative in the cocoon market indicates that bulk purchases are scheduled for that day and the demand shoots up taking along the prices too. This has also been one of the reasons that the government continues to operate these loss prone filatures.

The fluctuations in cocoon price and quantity are influenced by many factors such as cocoon crop yield, cocoon quality, season and demand for cocoons and silk. Thus, the yield and productivity from the cocoons do vary with season, which influences the economics of reeling operation. With fluctuations in cocoon prices and their quantities, sometimes the reelers are forced to buy cocoons at prices so high that the reeling activity may cease to be viable. Even then the reeling activity is continued in a fewunits just to ensure that the skilled workers, for whom considerable sums of money have been paid as advance, are retained.

The reeler is always short of working capital. Many reelers operate hand to mouth with their own money. They buy cocoons sufficient for a day or two and encash on their produce in the same frequency. They have little choice in terms of when to buy/sell or how much to buy/sell. They cannot cash on opportunity purchases and sales. Also, if the quantity of silk to be sold is small, the reelers' power to bargain reduces further and they may even sell it to a bigger reeler and not directly to the user. Very few reelers enjoy working capital support from commercial banks and generally the banks view the reeling sector as not worthy of credit support. Reelers have also revealed that their working capital requirements get almost doubled when the cocoon prices increase, while the profit margin remains the same.

The operations of the reeling units tend to be seasonal though there are no valid reasons to declare the reeling industry as a seasonal industry. The reelers make it seasonal largely due to financial compulsions. A few reelers, however, effectively operate the reeling units throughout the year.

The reeling activity has turned out to be more a gamble of making the right purchase of cocoons at the right time and selling the silk at an equally right time with little control on the environment, keeping the fingers crossed to see if it is going to be a profit at that point of time. Hence, study of economics over a period of time becomes more relevant than at any point of time.

Since the sector is unorganised and does not have vertical as well as horizontal integration, the reeler does not know about the requirements of his buyers and also as to who really needs quality raw silk. There are no opportunities for experience sharing or for evolving a collective approach to organise the industry structure. Self- serving interests are the order of the industry.

The raw silk price fluctuation follows the cocoon price fluctuation, which is in turn largely dictated by its supply. If the arrival of cocoons in the cocoon market improves, the raw silk price quickly comes down, while the previous stock of cocoons and silk might still be present with the reeler. This is one of the problems expressed by the reelers. They also express their displeasure about the absence of competition when they go to silk exchange, while they buy cocoons amidst tough competition from fellow reelers.

The fluctuation in the raw silk price may not have one to one correspondence with cocoon price fluctuation. The fluctuations in the raw silk price are dictated by the cocoon availability and its price. It can also be influenced by the availability and price of Chinese raw silk, festival or the demand season.

The reputation of the individual reeling clusters also has an influence on the price of raw silk. A few clusters like Sidlaghatta enjoy better price for raw silk as there is a perception that the quality of raw silk from that cluster is better. New areas and clusters with relatively less reputation face problems of marketing their silk at better prices. There is a covert demand for good quality silk and an overt demand for bad quality silk.

Quality weavers depend only on Chinese silk, which is uniformly of better quality and is easily available in large quantities at competitive prices. Handloom weavers, powerloom weavers and zari manufacturers across the country consume raw silk. Varanasi weaving cluster demands better quality finer denier raw silk and it is often a matter of pride for a reeler to say that his silk is sent to Varanasi or Surat. Varanasi and Surat traders pay a higher price for the silk but with a credit period of 3 to 4 months. The reelers who supply to these places are the ones with good financial background and therefore the waiting period is of no consequence to them. Thus they can realise better returns. More than the weavers, traders who have a better buying capacity, purchase raw silk in larger quantities and make a second sale to the weavers, either after twisting or as it is. Whereas, small time weavers, especially the handloom weavers, may not be in a position to purchase silk directly from the silk exchange since their requirements are small and they normally operate with credit facility or on job work.

As per the 1979 amendment of the Mysore Silkworm Seed and Cocoon Act, 1959, the sale of silk outside the state government exchanges is prohibited. Once the reeler enters the Bangalore silk exchange, he is not permitted to take back his silk without selling. The reeler has the option of selling or pledging for a later sale.

Silk exchange cannot ensure good prices, Karnataka Silk Marketing Board (KSMB) does not make regular purchases and traders do not participate in good numbers. In this situation, for a reeler who does not have adequate links in the trade to market his yarn finds it difficult to survive. However, unauthorised transactions are rampant and it is often quoted that less than 20% of the raw silk produced in the state is transacted in the silk exchange.

One of the main reasons for traders not purchasing at the silk exchange is the fear that their transaction would become legal and they will have to pay huge amounts of tax on their turnover, which happens to be quite high in this capital intensive business. The reelers feel that the prices of raw silk transacted outside the silk exchange are higher. The reelers complain that KSMB does not serve the purpose for which it has been established. The general feeling is that KSMB makes purchases when they have indents from their buyers and this obviously will be during high demand seasons associated with higher prices. They stay away from the market when the demand and thus the prices are lower, which is contrary to what is expected of them. Also, the contention of the reelers is that KSMB does not buy good quality silk since it is profit oriented and in turn has problems in selling better quality silk. This would have a negative influence on the trade.

KSMB would thus become indirectly instrumental in bringing down the prices since it participates in purchasing lower quality silk and thereby encourage the production of lower quality silk. Absence of a drive to produce better quality raw silk is attributed to lack of demand pull and non-availability of good quality cocoons in bulk quantity, consistently through out the year. Reelers often complain that there is neither demand nor appreciation for good quality raw silk.

The key to a filature's success is its supply of cocoons and the key to the quality of the cocoons are disease-free eggs. Weavers complain about non-availability of good quality raw silk in sufficient quantity throughout the year - the advantage they otherwise get from Chinese silk. The weaver also needs to be wary of consumer indifference.

Better reeling process would imply marginal decrease in production and justifiable increase in waste and the obvious improvement in raw silk quality. Most reelers refuse to acknowledge that this would pay in the long run. The other dimension of problem in quality raw silk production is that there does not exist a direct linkage between reelers and weavers, while traders play a vital role. This perhaps has adversely affected quality promotion since the user perception or demand for quality is not directly made known to the producer and this gap continues to exist. If this state of affair continues for a longer time, the future of sericulture industry would be in jeopardy and the absence of quality can be the single good reason for the decline of this labour intensive industry.

Better quality raw silk means higher cost of production, which may or may not be valued by the buyer, which may result in a loss to the reeler. Some reelers feel that, under the given circumstances, unless proper emphasis is laid on cost control, improvement in the quality of raw silk cannot be focused upon. The market situation does not support quality silk production, more so in the absence of quality based price fixation.

Not all reelers perceive the import of Chinese silk as a threat to the reeling activity-Some reelers feel that in the absence of Chinese silk, the demand supply gap would widen and prices would shoot up, which the weavers may not afford and industry as a whole may face problems. It appears that many of the problems expressed by the reelers are the banes of the industry in reality.

The development of any industry largely rests on the support it provides for quality improvement and sustenance. The system that exists does not provide means for encouraging quality cocoon production. Testing and grading procedure for cocoons and raw silk has not yet been standardised in India and prices are purely led by market situation. Quality based price fixation is the surest way of providing incentive to the producers and manufacturers to produce better quality.

Inspite of the research initiatives leading to better quality cocoons and better reeling technology, the industry has not benefited in real terms. There is no substantial improvement either in raw silk quality or in the financial status of the industry. Inferior cocoons at relatively high prices, poor raw silk quality, very low value addition, lack of appreciation for good quality indigenous silk and the like are some of the important causal factors.

Reelers continue to struggle to make both ends meet and the weavers have problems of sourcing and marketing. Thus there has been a vicious circle that is difficult to break. It appears that the various interventions of the government have created a 'House' but failed to transform it, into a 'Home'. Therefore, it can be seen that the silk reeling sector is plagued with many problems coming in the way of overall development of Indian silk industry.

The dialectics of our industry are far from answered. Is the policy of allowing import of Chinese silk in the interest of or against the industry? Is technology the right answer for improvement? Are subsidies the right approach for development? Has there been too much government support, or is the governmental presence in the wrong places? Are the market situations the most conducive for near perfect competition? How important is the quality aspect, technologically and commercially? Can the reeling unit ever thrive under the given situation and if it can, what are the pre-requisites? Is auctioning the right approach for transaction? Is the undue reliance on subjective assessment of quality by the buyers, in the interest of growth of the industry? Will quality based pricing be a better approach for the good health of the industry? These kinds of questions can go on and on. Should not an industry with such huge stakes economically, socially, technologically and of course culturally, deserve a much better treatment?

A detailed analysis of the situation as it exists is required to bring to light the factors responsible for the present status. Once the causes are identified and quantified, the approach for remedying the ills can be arrived at. There are adversaries at both the ends viz., cocoon production on the one hand and fabric production on the other and the reeling industry is caught between these problem areas, in addition to its own inherent problems.

The need of the day is to deliver better quality at lower cost. Cost cannot be disassociated from quality and thus lower cost becomes a quality parameter. The possibility of reducing the cost of production by more efficient processes and passing on this advantage to the consumers would become one's competitive edge in the market.

Attempts have been made earlier to study the impact of certain new advancements in technology or financial assistance made available in the form of loans to the reeling industry. But such studies have in practice been conspicuously unrealistic, assuming for instance a regularity of operation contrary to the basic characteristics of the industry.

According to Charsiey, such studies do not appear to rest on any direct study of those engaged in commercial production but rather on the application of notional figures to a largely a priori model. Even when the primary data have been collected from the commercial units in a few studies, the problem has been that the data pertains to a point of time and does not, therefore, reflect the fluctuations in the operations, which in fact decide the fate of the reeling enterprises.

Concern with how the real business operates in practice over a period of time is therefore very important. Studies on the economics of reeling operations done in the past indicate negative net revenue for the reeling units. However, if the reeling industry has survived over the years, it cannot be in spite of the loss. The industry has to breakeven over a period of time to consolidate the gains, however meagre. So it is worthwhile to study the economics of silk reeling industry over a sufficient period of time that is enough to include a cycle of seasonality, be it with respect to cocoon quality, cocoon availability, cocoon price, raw silk demand or raw silk price.

Since reeling activity is highly input dependent, ignoring the aspects of input market conditions would not yield a holistic perspective of the operations. If one has to study and understand the market dynamics, the attributes of the commodity transacted should be clearly known. The commodity in question here is cocoons and the attributes are its price, quantity and quality.

Cocoon prices are explicitly available and the supply i.e.. total quantity of cocoons transacted on a given day is comprehendible from the cocoon markets. However, the quality of cocoons does not have a unique expression, either to consider as a variable for studying the relationships in the market or as an input into the production process or as a yardstick for commercial transactions.

Cocoon quality influences reeling productivity, raw silk yield and quality and thus in turn the profitability of reeling operations. It may be noted that, variations in cocoon price over a period of time are dictated by the seasonally with respect to cocoon supply and demand. With regard to variations in cocoon price at a given point of time (within a season), price differential between the cocoon lots appears to be largely due to its quality.

Study of the entire industry as a system would result in aggregation, thereby truncating the vital details that would be more meaningful and realistic. Nonavailability of reliable data was yet another constraint for studying the silk industry as a system. The other useful alternative is to study the operations of a typical reeling unit to gain an insight into the intricacies of its day to day operational problems and the aspect of decision making in the light of the volatile situation in the industry.

5

Raw Silk Reeling

Silk reeling is the process by which a number of cocoon baves are reeled together to produce a single thread. This is achieved by unwinding filaments collectively from a group of cooked cocoons at one end in a warm water bath and winding the resultant thread onto a fast moving reel. Raw silk reeling may be classified by direct reeling method on a standard sized reel, indirect method of reeling on small reels, and the transfer of reeled silk from small reels onto standard sized reels on a re-reeling machine. The last technique is primarily applied in modern silk reeling processes.

Types of Silk Reeling Devices

There are many types of silk reeling machines in use. The major structural features of the Sitting Type Reeling Machine, the Multi-ends Reeling Machine and the Automatic Reeling Machine are shown in Table 1.

Table 1. Comparisons of structural features of various reeling machines

Kinds of machine	*Sitting type reeling (m/c)*	*Multi-ends reeling (m/c)*	*Automatic reeling (m/c)*
No. of reeling ends per basin	4 (2 to 8)	20 (10 to 40)	20

Contd....

Contd....

End-groping apparatus	Hand-driven	Semi-automatic	Automatic
End-picking apparatus	Hand-driven	Hand-driven	Automatic
Cocoon supplying apparatus	None	None	Equipped
End-feeding apparatus	None	Equipped (hand feeding)	Equipped (machine feeding)
Stop motion	None	Equipped	Equipped
Traverse guider	Equipped	Equipped	Equipped
Temperature of reeling bath	65-80° C	30-45° C	30-45° C
Reeling velocity (meter per minute)	180-250	50-80	120-200

Hand Spinning Wheel

This primitive spinning apparatus is operated by two hands—one to drive the wheel and the other to feed in cocoons. One end of the reeling thread is wound onto each wheel, while cocoons are boiled in a separate pot.

Charka Type Reeling Machine

The Charka type is in use in India. This machine is operated with separate work motions in reel driving and cocoon feeding to reeling ends by two men per machine. Each machine has 3 ends or more to a reel, which is the same size as the large wheel of the Re-reeling machine in order to save the re-reeling process (direct reeling method).

Sitting Type Reeling Machine

There are two kinds of sitting type reeling machines, foot operated and motor-driven. The motor-driven reeling machine is not equipped with the stop motion attachment. There are obstacles to the production of good quality raw silk as the raw silk thread is wound too rapidly to maintain good quality control.

Multi-ends Reeling Machine

This machine eliminates the disadvantages of the Sitting type reeling machine by increasing the number of reeling thread ends per basin and reducing the reeling speed. The operator must stand when running this machine as the number of reeling threads per basin increases by twenty-fold. This is also called a "Standing type reeling machine". Reeling efficiency is unchanged. Quality is better due to reduced speed.

The Multi-ends reeling machine is composed of driving part, groping ends, picking ends, standby bath, reeling part, jetboute, stop motion, traverse guider, small reels, steam heating pipes and clutches.

The cooked cocoons contained in the tubs are carried into the groping ends portion of the reeling machine. From there, cocoons are moved into the picking ends apparatus. After correctly processing, the cocoons go to the standby bath for cocoon feeding. They are picked up by the reeler and fed to the reeling thread. During this step a number of cocoons will be dropped thus reducing the ratio of reeling cocoons per thread. The normal speed of cocoon feeding by a skilled reeler is around 16 times per minute. The reeling thread passes through the jetboute, silk button, first guider, second guider, third guider, fourth guider, traverse guider, in that order and then is wound onto the small reels.

The cocoons dropped during the reeling process are gathered and reprocessed starting from the groping end section. The croissure of reeling thread is made between second guider and third guider, and the length of croissure is not for twisting of thread but for cohesion of thread by rubbing of composed filament. Typically, one set of Multi-ends reeling machines consists of ten basins with each basin having twenty ends or reels.

Basin: The basin is rectangular with well-rounded corners and edges. It is only 10 to 12 cm deep. It is commonly made of dark coloured porcelain. The basin is subdivided into sections, each intended for a specific job such as brushing, end gathering of baves, stocks in reserve and waste collection.

Reels: The reels of the Multi-ends reeling machine have a circumference of 75 cm. The frame of the reel is made of light metal or plastic. The reels are fitted into reel carriers and driven by a transmission shaft by connecting gears.

Traverse guider: To ensure narrow and long web on the hank of the reel, a cam type traverse assembly has been fixed. This will make a convex surface in the hank, which is wound on the reel. The centre part of the hank is higher than the two axis.

Thread button: Porcelain button thread-guiders are used for removing any dirt adhering to the thread passing through the tiny aperture in the button.

Automatic Reeling Machine

In raw silk production, the continuing increase of labour costs has mandated automation. Around 1950, the Automatic reeling machine, which controls the number of reeling cocoons per thread, was invented. Shortly thereafter, it was replaced by a second Automatic reeling machine, which could automatically control the size of the reeling thread.

The Automatic reeling machine mechanizes the processes of groping ends, picking ends; cocoon feeding to reeling thread and separation of dropped end cocoons during the reeling process. The efficiency of the Automatic reeling machine compares favourable with the manual Mult-ends reeling machine.

The Automatic reeling machine though built to replace manual reeling, still requires manpower for problems with the reeling thread, which must be corrected by hand. A moderate amount of cooked cocoons are carried to the newly cooked cocoon feeder (3) and then removed into the groping end part (4)

The end groped cocoons go to the picking end part (5) and the correctly picked end cocoons are dispensed to the cocoon supplying basket (8) which continuously rotates around the reeling basin on an endless chain belt. Usually, the reeling method is classified into the fixed cocoon feeding system and moving cocoon-feeding system.

In the case of the fixed cocoon feeding system, the correctly picked end cocoons in the rotating cocoon baskets are poured into the arranging basin (11) and here the picked end of each cocoon is hung on the end holding reel (7). When the size detector of the reeling thread indicates the feeding of cocoons, the picked end cocoons on standby are fed to the reeling thread by a feeding spoon. The reeling thread fed by picked end cocoons passes through the jetboute (23), silk button, first guider, second guider, third guider, fourth guider, denier indicator (5′), fifth guider and traverser (3′), and then it is finally wound onto small reels (1′). The end dropped cocoons are placed into the cocoon flowing tunnel by the remover plate. They are carried into the pupa separating drum (14).

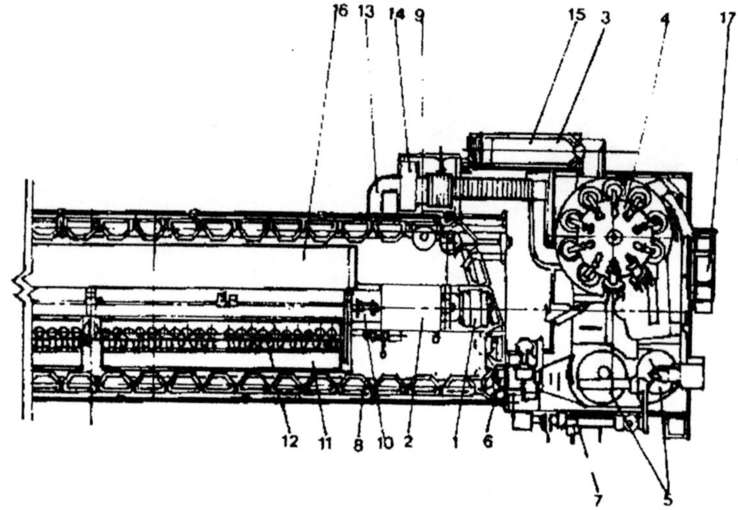

Figure 1. Ground plan of automatic reeling machine

However, more reelable cocoons are poured into the end groping part (4) by the conveyor belt (15) and reels-finished cocoons are placed into the dropped-pupa case (17) for parchment layer cocoons (Figures 1 and 2).

In the case of the moving cocoon feeding systems, the correctly picked end cocoons are contained in the moving cocoon basket equipped with cocoon feeding apparatus. They are fed by the feeding fork of the cocoon basket, which move simultaneously around the reeling basin. The denier indicator of the reeling thread indicates the feeding motion of the cocoon. After cocoon feeding, the reeling path of the moving cocoon feeding system is the same as that of the fixed cocoon feeding system.

Generally, one set of the Automatic reeling machines has 400 ends, while one basin has 20 ends. The operating efficiency of the Automatic reeling machine is easily affected by cocoon qualities, drying and cooking machinery and quality of reeling water.

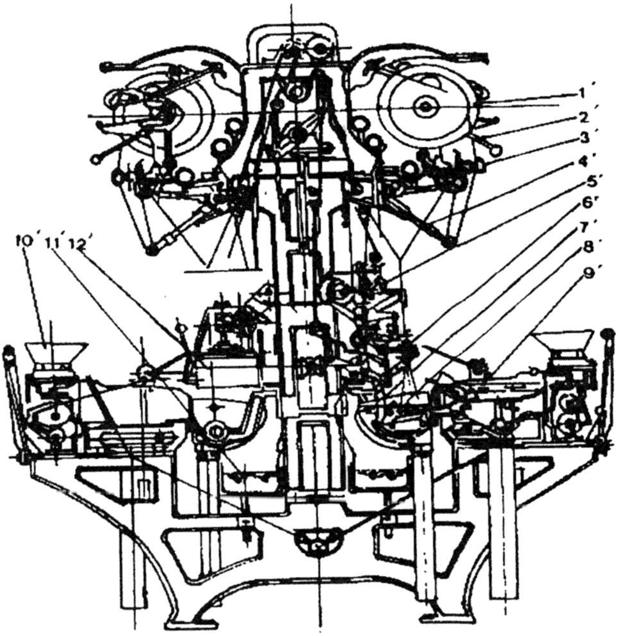

Figure 2. Side view of automatic reeling section1'. Small reel, 2'. Rotating device, 3'. Traverser, 4'. Stop motion, 5'. Denier indicator, 6'. Jetboute, 7'. Shatter for reeling cocoons, 8'. Reeling basin, 9'. Arranging basin, 10'. Cocoon supplying basket, 11'. Conveyor gutter, 12'. Rotating device for jetboute.

Brushing and Picking Section

An automatic ten-brush unit brushes cocoons with independent arms rotating in reverse motion on an axis in circular basins.

Picking frames rotating in one direction pick off the brushed cocoon filaments when a cam during the operation raises the brushes. Selective picking is completed in the most effective manner by this equipment. To maintain the exact number of cocoons at each reeling end, a control device is attached which detects the amount of cocoons in the cocoon suppliers and automatically supplements deficiencies in the number of cocoons.

Reeling Section

The denier indicating and detecting device

In these devices, the yarn constantly passes through the denier indicator and detector, which are set to a given size. The size of thread being reeled is detected through the balance between the friction of the running thread and an eccentric weight fitted on the denier indicator.

When the thread becomes thinner than the fixed limit, the denier indicator indicates the necessity of feeding-ends. This indicator is composed of two gauge glass plates with other pieces of gauge plate which have a slit between them corresponding to the given size of thread to be reeled. The size of yarn may be adjusted to the required sizes by varying the irregular weight with the denier-adjusting device; if a wider range of adjustment is required the denier indicators have to be replaced.

Conveyor system for cocoon suppliers

The fixed end feeding system is employed together with a conveyor system to carry cocoon supplies and feed cocoons to the reeling section whenever required.

End feeding for cocoons suppliers

Cocoons are supplied to the conveyor, which rotates constantly around the reeling section. The feeding lever fixed to the detecting mechanism will trigger the driving lever on the cocoon supplier only when the size of yarn becomes thinner than the required denier during reeling operation.

Stop motion

If there is a defect in the reeling, the reel is automatically stopped by a brake, which is activated by contact pressure from operation of the detector level.

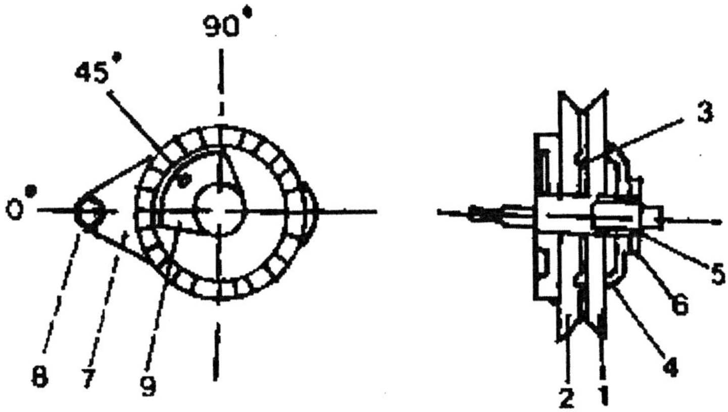

Figure 3. Size detector (gauge type) 1. Gauge glass, 2. Grooved gauge glass, 3. Gauge plate, 4. Washer, 5. Binding bolt, 6. Special nut, 7. Indicating ring, 8. Detection limiting rod, 9. Adjusting weight

Dropped cocoon gatherers

The apparatus gathers baskets that have collected all cocoons dropped during end feeding. These are carried to the dropped cocoon separator. These baskets travel intermittently between the cocoon suppliers.

Separator of dropped cocoons

The device accurately distinguishes and separates pupae, dropped middle layer cocoons and thick layer dropped cocoons.

Semi-automatic Reeling Machine

This is a mixed-mode or hybrid machine between the automatic and multi-ends reeling machines. As an application, this type is better for improved reeling efficiency and raw silk quality than the multi-ends reeling machine. The semi automatic reeling machine can be operated with poor quality cocoons, but relies on more labour than the automatic reeling machine.

The cooked cocoons are carried into the groping end part (10) and the end-groped cocoons are removed to the picking end part (11). The correctly picked end cocoons go through the cocoon supplying basket (9) and then to the arranging basin (8). When the size detector of the reeling thread indicates the feeding motion, the picked end cocoons on standby are fed to the reeling thread by a feeding spoon. The reeling thread passes through the jetboute, silk button, first guider, second guider, third guider, fourth guider, size detector, fifth guider and traverse guide, and then it is finally wound onto small reels (4) (Figure 4).

The end dropped cocoons are collected by the cocoon buckets (13) and removed to the cropped cocoon basin, where more reelable cocoons are separated and then poured into the groping end part (10). The principal difference between a semi-automatic reeling machine and an automatic reeling machine is that the cocoon end groping, cocoon end picking and cocoon carrying are manual.

Methods of Silk Reeling

Reeling Velocity (R.V.)

Reeling velocity is defined as the winding speed of raw silk on the reel. The reeling velocity is measured as the length of raw silk reeled during 1 minute or as the revolution number of the reel per minute. It is usually calculated by the following formulas.

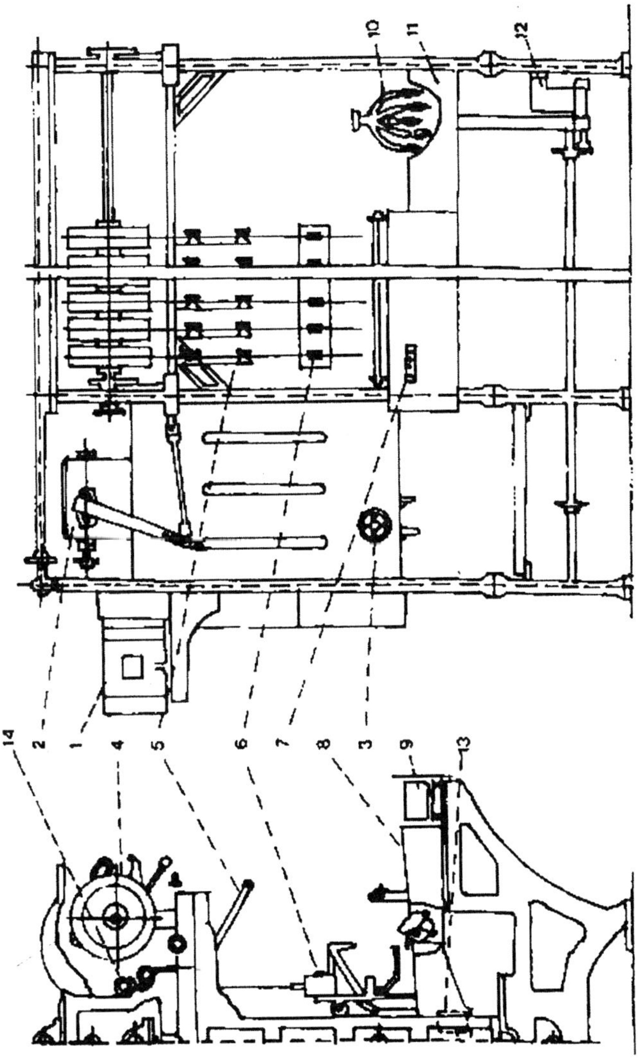

Figure 4. Dimension and structure of semi-automatic reeling machine1. Motor, 2. Traverser gear, 3. Chain stepless regulator, 4. Small reel, 5. Stop motion device, 6. Denier control device, 7 Brush, 8. Arranging basin, 9. Cocoon carrier, 10. Brush, 11. Picking bath, 12. Brush drive device, 13. Reconveying bucket (dropped cocoon), 14. Steam pipe.

$$\text{R.V.(m / min.)} = \frac{\text{None breaking reelable bave (m)} \times \text{number of end feeding bave (pc / min/ reel)}}{\text{Ave. reeling cocoon number per thread}} \qquad (1)$$

$$\text{(1) R.V.(m / min.)} = \frac{\text{None breaking reelable bave (m)} \times \text{number of end reeling (pc / min.man)}}{\text{Ave. reeling cocoon number per thread} \times 20 \text{ (reels)}} \qquad (2)$$

Where formula (1) applies to the automatic type, formula (2) applies to the multi-ends type.

Number of end feeding :

Automatic type—1.2-1.5 pc/min./reel

Multi-ends type—12-14 pc/min./man

Ave. reeling cocoon number per thread =

$$\frac{\text{required size of raw silk}}{\text{size of cocoon filament}}$$

None breaking reelable bave =

length of cocoon filament (m) x cocoon reelability (percent)

The velocity of cocoon reeling affects the raw silk yield, reeling efficiency and raw silk quality (Table 2 and 3). Optimum reeling velocity should be maintained for the best possible product.

Groping Ends

The high water temperature of the groping ends section dissolves sericin to a great extent and improves the reelability of the cocoon. However, it is apt to reduce raw silk percentage of the cocoon. Alternatively, too low water temperature in this process can reduce the groping end efficiency, the reelability percentage and the reeling efficiency.

Table 2. Effect of various reeling velocities on silk yield and quality with different grades of reelability cocoon

Cocoon reelability	Reeling velocity (m/min.)	Raw silk percent of cocoon	% of cocoon reelability	Raw silk quality	
				Neatness (%)	Cleanness (%)
Good Group	90	16.87	72	96.2	96.8
	120	16.80	71	95.2	98.8
	150	16.73	71	93.4	94.5
Fair Group	90	16.36	58	91.9	93.7
	120	16.32	57	90.0	93.8
	150	16.21	55	90.5	93.0
Poor Group	90	16.17	50	91.0	92.3
	120	16.07	49	89.5	90.8
	150	15.70	46	89.0	90.6

Table 3. Relationship between reeling velocity and evenness of raw silk quality

Reeling velocity (m/min.)	Average evenness (%)	Low evenness (5)
50	93.5	86.0
80	98.0	82.0
110	88.5	80.2
140	87.3	80.0
170	85.5	78.4

An appropriate temperature range is from 75°C to 80°C, but it will vary. The number of groping end cocoons is about 70-90, but this also varies with the cocoon quality. The pH value of water at this point should be from 6.8 to 7.0. the water should be changed often to prevent discolouration of raw silk.

Picking Ends

This is a very important function of the reeling operation, which can affect raw silk percentage and reelability. Correct end cocoons should not be permitted in picking ends to minimise cocoon filament waste. Insufficient picking ends causes the feeding of incorrect end cocoons, breaking down of the reeling thread, and affects the accuracy of the feeding end work.

On the other hand, excessive picking ends reduce raw silk by causing an increase in waste. The picking end with too many cocoons can cause the work to be carried out inaccurately and is apt to increase waste of cocoon filament by repeated work on correct end cocoons. The number of cocoons for a picking end section should be below 40 pcs, and the recommended temperature of the water bath is 40 to 50°C. In automatic end-picking machines, the degree of end picking varies between 80 to 85 percent for good cocoons, 75 to 80 percent for medium ones and 70 to 75 percent for inferior ones.

Reeling Bath

— *Temperature*—High temperature in the reeling bath increases the breakdown of the reeling thread, cleanness defects and reduces raw silk yield. It improves the reelability of cocoons, the cohesion and the feel of raw silk by reducing the gum in the sericin. The optimum reeling bath temperature depending on cocoon characteristics is about 30 to 45° C.

— *Consistency of water* — Continuous reeling work without changing of reeling water results in water, which is dense, coloured and acid. High density water reduces the cocoon reelability, but increases raw silk percentage making the silk light greyish in colour. Water supplied to the reeling bath should be pH controlled so that the water for each reeling bath may remain the same to ensure colour uniformity. The best consistency is when the pH of the water is about 6.8-7.0, and the water contains a little diluted and floating sericin.

— *Supply of cocoons* —In automatic reeling machines, cocoons are supplied in two ways: rotary type and fixed type. In both types, when the size detector is activated, cocoons are supplied one by one, accurately and quickly. In order to raise the feeding efficiency, it is important to minimise the end-missing cocoon supplier and eliminate the floating cocoons.

— *Maintenance of size detector* — The identification of size by the detector should be accurate. In practice, however, the performance of the detector can be badly affected by improper care, dirt, and abnormal resistance of thread passage and other defects. During operation, the detector should be kept clean and in good repair. Periodic inspections and maintenance should be carried out.

— *End feeding and button* — End-feeding should furnish the end within 1 to 2 seconds without long knots. The hole of the button should correspond to the size of the raw silk

The diameter of the buttonhole should preferably be in the following range:

$$39.0\sqrt{S} \langle D \langle 52.0 \sqrt{S}$$

Where S is the size of raw silk, and D is the diameter of botton hole.

— *Control for breakdown of a reeling thread*—Inefficient processing is considered to be the main cause of frequent breakdown, which in turn decrease reeling efficiency. The reeling velocity has to be controlled so that the total working period of repairing any breakdowns may be less than 60 percent of one pulse operation period. The breakdown ratio by snap reading is about 1.5 percent to under 2 percent.

— *Arrangement of reeling part*—Disordered reeling parts make reeling difficult, reducing the raw silk percentage and the reeling efficiency, while deteriorating the raw silk quality. Reeling conditions of reeled cocoons should be monitored through continuous inspection.

Quality Control During Raw Silk Reeling

The purpose of reeling process is not only to raise the raw silk yield of cocoons and reeling efficiency, but also to improve raw silk quality.

a) *Size development and evenness*—To maintain reeling thread in the required size, the average cocoon number per thread must be adjusted by a check to produce silk thread in the same size throughout all ends during reeling. If the size is different from the required size it should be readjusted by the group size controller. This will reduce size deviation. By improving the accuracy of cocoon supplying each silk thread becomes uniform in size and different skein sizes are reduced. It can also improve the size deviation and evenness of raw silk quality through accuracy of cocoon supplying work and the improvement of cocoon reelability.

b) *Defects* — Defects in raw silk are divided into super major defects, major defects and neatness defects. Defects occur based on reeling conditions. Types of defects and remedies are explained in Table 4.

Table 4. Prevention method against various types of cleanness defects

Types of defects	*Prevention methods*
Waste	1. Control of over or uneven cooking.
	2. Prevention of carry over cocoons by remaining in basins during reeling process.
	3. Reducing the number of standby cocoons in the reeling basin.
	4. Preventing the over swelling of cooked cocoons in the reeling process by controlling the temperature of grope end and picking end bath as well as maintaining a constant temperatur
Large slug	1. To 4) are same as above.
	2. Avoid low temperature cocoon drying
	3. Controlling the excessive increase of reeling speed and reeling thread
Bad casts	1. Keep the extra end of cocoon filament short when the correct end picked cocoons are fed to the reeling thread.
Long knots	1. When the broken thread are tied and knotted during the reeling or re-reeling process, the end of the tied and knotted thread must be cut off, leaving only 2 mm.
Cork screws	1. When the reeler has to eliminate a reeled out or bad cocoon during reeling, he

Contd....

Contd....

	should not use force to pull it out but should snip it out with his fingers as quickly as possible.
Split ends	1. Reduce friction effect on silk thread. Pulleys with cracks and splits on the surface should be removed. All pulleys should be perfectly polished to reduce split ends in the reeled yarn.
Long loops	1. Prevention of over or uneven cooking.

c) *Cohesion*—Good cohesion raw silk is needed for the warp of silk fabric. Factors that improve the cohesion are temperature, amount of reeling tension, sufficient croissure and good drying of raw silk.

d) *Colour*—The water consistency and temperature of groping end part and reeling part affect the colour of raw silk. Therefore, the temperature control and water supply in the reeling machine should be constantly monitored to obtain a uniform colour of raw silk.

6

Re-reeling

Re-reeling is a process of reeling the raw silk from small reels onto large (standard) reels with a circumference of 150 cm, adjusting the width and weight of skein uniformly. The object of re-reeling is to make standard sized hanks or skeins for marketing purposes. Skeins should have a circumference of 150 cm in accordance with International Standards. Secondly, re-reeling eliminates defects, which might have occurred during reeling.

Re-reeling Machines

Basically the re-reeling machine consists of various parts and instruments such as the grand reel, reel revolution and stop motion equipment, traverse bar and guider and drying pipe, double thread checker and accessory utensil (Figure 2). The circumference of the grand reel is 1,50 cm and length of the reel rib to be used for five skeins is usually 0.67 meters.

Process of Re-reeling

Pre-treatment

To obtain best results from the re-reeling process, the sericin gummed at the rib edge of the small reel should be swollen and softened through a dipping process.

i) Infiltration method by vacuum tank 203 times infiltration with 30-35° C water under 40-60 cm Hg is recommended (Figure 1).

Figure 1. Vacuum pre-treatment for rereeling

Figure 2. Raw silk rereeling machine

ii) Bath method. Dipping time normally 10-30 minutes, but should be changed depending upon the

temperature and concentration of the chemicals used. Currently emulsified oil is recommended for use as a chemical for re-reeling.

iii) Damping method. It is sometimes desirable that the hard gummed raw silk on the small reel be covered with a wet cloth.

Dipping

In the case of a double skein, additional water is occasionally needed when there has been inadequate wetting of the raw silk on the small reel. This is accomplished by use of water spray, dipping and wet cloth, for instance.

Re-reeling Velocity

There is a tendency that re-reeling velocity is closely related to the number of breaks of the raw silk, re-reeling efficiency and generation of the gummed skein. Optimum velocity recommended for re-reeling is 150-170 rev./min. for 21-denier and 130-160 rev./min. for 14-denier raw silk.

Table 1. Temperature and humidity of re-reeling in the machine

Temperature (° C)	*Relative humidity*	*Moisture content of raw silk skein after re-reeling (%)*	*Appearance of raw silk skein*
31	55	9.7	Gummed
41	36	6.2	Normal
68	29	2.7	Over dried

The best temperature and relative humidity of re-reeling is around 41° C, 36 percent R.H. Under this condition the moisture content of raw silk skein immediately after re-reeling reaches 6.2 percent and the raw silk skeins are not stuck together (Table 1).

Raw silk should have about 6-8 percent moisture in it after the re-reeling process is finished. The number of grand reels per labourer should be around 15-20 grand reels for 21-denier and 20 grand reels for 14-denier. This will change if the raw silk size, re-reeling velocity and number of thread breaks are not within expected bounds.

Re-reeling Efficiency

The standard of the re-reeling velocity (Re.V.) and the re-reelable quantity of the raw silk can be calculated according to the following formula.

$$\text{Re.V. (rev.Min.)} = \frac{\text{weight of skein x 9000} \div \text{size of raw silk (d)}}{\text{re} - \text{reeling hour}}$$

$$\times \text{ workable number of grand reel by a labourer} \times 1.5$$

For example, $130 \times 9\ 000 \div 20.6/8 \times 30 \times 1.5 = 158$ rev./min.

For the calculation of the re-reelable quantity of raw silk by a grand reel for 8 hours, the following formula should be used.

$$= \text{Size of raw silk (denier)} \times \text{re} - \text{reeling velocity (rev./min.)} \times 1.5\text{ m}$$
$$\times (0.05 \div 450) \times 5 \text{ skeins} \times 480 \text{ min.}$$
$$\times \text{ effective working ratio of re} - \text{reeling}$$
$$= \text{Size of raw silk} \times \text{re} - \text{reeling velocity} \times 0.4$$
$$\times \text{ effective working ratio of re} - \text{reeling (Normally 85\%)}$$

For example, $20.6 \times 170 \times 0.4 \times 0.85 = 1.19$ kg.

Silk End Tying and Skein Lacing

The tie end of the raw silk should be tied with cotton year, to find the silk thread end easily during the preparation process for weaving.

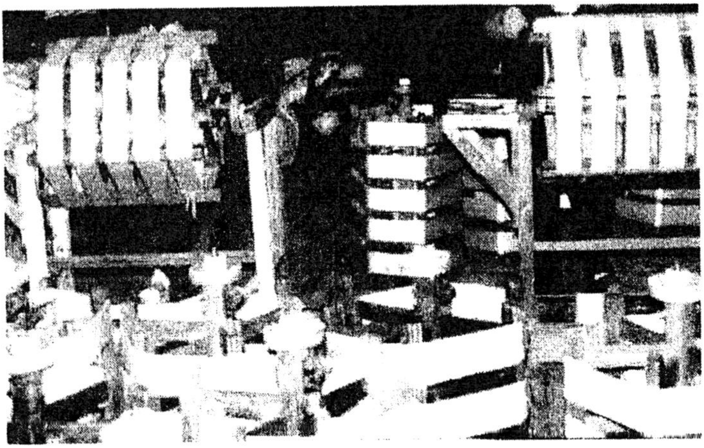

Figure 3. Lacing of silk skeins

This is known as silk end tying. Skein lacing keeps the diamond cross originally in the skein place. Five interlacings each at three places on the skein are suitable to prevent tangling (Figure 3).

Booking and Packing

In order to provide proper moisture to the raw silk, the skein should be kept between 20-23°C and 65 to 75 percent relative humidity for several hours prior to booking of the skein.

Standard of skein size: There are two standard skeins, the 70g known as a single skein and the 125-135g size known as a double skein. The double skein is the preferred version.

Booking of the skeins: With the bleached cotton rope doubled with 74 plies of 10s yearn, the 30 skeins must be bundled into a book if they are single skeins (5 line 6 stairs). In the case of double skeins, a book consists of 20 skeins.

Packing of a lot: One bale contains 22-30 books and the standard weight is about 60 kg.

Storage of Silk

Raw silk is vulnerable to heat and sunlight, which may alter its colour, while moisture may cause fungal attack. Thus, precautions must be taken to store raw silk. The storeroom must maintain control over temperature and humidity. BHC powder, chloropicrin, naphthalene and other repellents should be utilised to counter damage from insects. Open skeins should not be stored over long periods; properly packed books and bales can be stored for extended periods under strict atmospheric conditions with no deterioration to the raw silk quality.

7

Use of By-products in Silk Reeling

In addition to the chief products of the sericultural industry, there are a host of waste materials eliminated during manufacturing, which have commercial value During rearing, pierced cocoons and double cocoons are created. Silk reeling generates brushing waste, end-missing cocoons and pupae. Double cocoons are processed to make doupion silk. Other rejected cocoons and waste are re-processed into flow silk and spun silk yarns. The pupae are sold for fertilizer, feed stuff and other agricultural purposes.

Doupion Silk Reeling

Natural fancy doupion silk can be made from double cocoons. Current practices also mix inferiors single cocoons with double cocoons in the reeling process (Table 1). Coarse size 225d or 110d doupion silk is reeled by one end only; finer 60d doupion is usually reeled by 2 ends. Reeling parameters are velocity of 230 m/min and the reeling bath temperature of 80-90°C. The reeler gropes the cocoons in the pot with the groping broom held in the right hand and feeds the ends to the thread passing through the guide.

Reeling of Non-mulberry Cocoons

Tasar cocoon reeling

Tasar cocoons have a compact structure and composition distinct from that of mulberry cocoons. The cooking methods and chemical treatment shown in Table 2 reflect details of Tasar processing. The Sitting type doupion silk reeling machines are used for 110 and 225 denier Tasar silk, and Mult-ends reeling machines or Pedal reeling machines are used for the fine 42-63 denier silk.

Table 1. Reeling efficiency of doupion silk by the sizes of thread

Sizes of thread	*Formation of raw material cocoons*		*Doupion silk (%) of cocoons*	*Reeling efficiency (gams per person in 8*
	Double cocoons	*Inferior cocoons*		
225 denier	50%	50%	14.2%	1,006
110 denier	50%	50%	14.3%	659
60 denier	50%	50%	14.0%	459

Table 2. A trial of Tasar cocoon reeling

Cooking			
First stage	*Second stage*	*Reeling*	*Results*
A. Solution (alum 0.2% formalin 0.05% nonionic surfactant 0.03%) Boiling for 30 min.	Solution (NaOH 0.1%, H_2O_2 0.2% nonionic surfactant 0.03% Soaking for 24 hrs.	Sitting type reeling machine (for doupion silk reeling) Reeling cocoon number per thread : 40 pcs.	Size of thread (denier) 225 Weight of reeled silk per reeler : 125.6 gr./1 hr. Silk ratio of cocoon : 8.25%.

The "Natwa" pedal reeling machine used for fine Tasar reeling is made of bamboo and wood. While productivity is low, it is widely used in India because it is simple to operate and requires small investment cost.

The Pedal reeling machine is derived from spinning equipment with twisting and winding of the yarn without a fan system. The machine has 4 spindles and a wooden wheel of 50-cm circumference for winding the yarn. It is driven by a foot pedal and the cocoons are reeled by hand. Preferred sizes of Tasar silk are 40/44d and 60/66d.

Eri cocoon spinning

Eri cocoons are open mouthed with a discontinuous filament, which make them suitable only for spinning. Approximately 90 percent of Eri cocoons are hand spun in Assam, India. The characteristics of Eri cocoon bave are shown in Table 3.

Table 3. A survey of characteristics of Eri cocoon bave

Cocoon	*Cocoon bave*	*Composition of bave*
Weight (g) : 2.4-2.6	Size (denier) : 1.77	Fibroin (%) : 72.2
Shell ratio (%) : 13	Tenacity (g/d) : 3.5	Sericin (%) : 11.9
Length (mm) : 3.5-5.5	Elongation (%) : 20.8	Fat (%) : 1.3
Width (mm) : 14-25		Moisture (%) : 14.6

Manufacture of Floss Silk

Floss silk may be manufactured from any kind of cocoons, but principally it is processed from pierced, end-missing, and double cocoons. Floss silk is beneficial as paddy against cold weather and as a basis for hand spun yarns. The procedure to create floss silk involves degumming, opening-up and finishing.

Degumming

Cocoons are boiled in hot water with added alkali to remove gum and make them soft. If degumming is insufficient, the opening-up will not be easy: conversely over degumming makes floss silk fragile. An example of chemicals used:

NaOH	: 0.05%
$NaHCO_3$	: 0.2%
Surfactants	: 0.03%
Boiling time	: 70 min.

Opening-up

Degummed cocoons are put in clean water and opened up individually by turning each one inside out to remove pupae and other waste. Several cocoon shells are stacked to form a bog into which the fingers of both hands are placed and spread in the water. These are then hung on a frame. Generally, when a sheet of floss silk reaches around 6 grams, it is taken from the frame.

Finishing

When the floss silk is removed from the frame, it is either soaked in clean water or a 1% hydrogen peroxide solution for a few hours to refine the colour. The, the liquid is squeezed out and the floss silk is shaped to be air dried in the shade from a pole or a rope.

Spun Silk

Short lengths of inferior silk filaments taken from waste material are combed and spun together as silk thread. Spun silk threads are soft but less lustrous, strong and elastic than reeled silks. Spun silk fabric will become fuzzy with wear as the yarn is made from short material.

There are several sources to fabricate spun silk:

1. Pierced cocoons, the result of breeding moths that have emerged from their cocoons;
2. Double cocoons or doupion, which result when two cocoons have been spun too closely together;
3. Floss, brushed from cocoons before reeling;
4. Friese, the coarse and uneven silk fibre at the beginning and end of each cocoon;
5. Scrap, the machine waste left over from reeling.

The spun silk manufacturing processes are as follows:

Silk wastes → Degumming → Washing → Dehydration
→ Drying → Beating → Wetting → Opening
→ Filling → Circular Dressing → Spreading → Settling → Drawing →
→ Combing → Roving → Fine Combing → Drawing
→ Slubbing → Re-spinning → Fine spinning → Doubling
→ Twisting → Gassing → Reeling → Spun Silk

Figure 1. Spun silk manufacturing process

Spun silk is less expensive than reeled silk. Although spun silk has less strength and elasticity than reeled silk because of the shorter staple used, it possesses all the general characteristics of reeled silk. Tub silk fabric, for example, is made of spun silk, yet it gives good service when the quality of the fibre is good. Spun silk is used for shantung, pile fabrics, dress trimmings and linings, elastic webbing, sewing silk, summer weight silks, velvets, umbrella fabrics and insulation.

The waste derived from the processing of spun silk yearn is also used. Such fibre is labelled as waste silk, silk waste or most frequently noil silk and silk noil. Silk noil may be reprocessed into spun yarn ad woven into textured fabrics for draperies, upholstery and sportswear. These fabrics are dull, rough, with a cotton-like appearance and more resilience.

Yarn Count of Spun Silk

The size of spun silk thread is defined in a similar manner to standards used for cotton yarn. For cotton, the term "2/60s"signifies a two-ply yarn consisting of two single strands twisted together, each having a yearn count of 60.

In the case of spun silk the notation has a different meaning. For example, for 60/2 two yearns with a separate yarn count of 120 have been doubled, producing a ply yarn with a new count of 60.

Wild Silk

Silkworms hatched from the moth Antheraea mulitt, live on oak leaves instead of mulberry leaves consumed by cultivated species. Oak leaves produce an irregular and coarse filament that is hard to bleach and hard to dye. Wild silk has a tan colour derived from the tannin in the oak leaves. It is commonly woven from this naturally coloured thread and is rarely dyed except in solid shades. Standard wild silk thread is made from eight cocoons and averages 32/34 denier. Wild silk is less lustrous than cultivated silk, since about 11 percent of sericin is removed in the degumming process. Wild silk fabrics are durable, washable and less expensive than pure dyed silk. Typical finished fabrics in this category are rajah, shantung, tussah and pongee.

8

Grainage Technology

Site Selection

Select grainage site in a well ventilated area, avoid establishing in areas with high temperature, near factories and on road side. Ensure adequate water supply to grainage.

Selection of Seed Cocoon

Select uniform cocoon in size, shape and texture. Avoid double cocoons. Reject seed cocoons with high floss contents and melted cocoon. Ensure cocoons having the following criteria for selection.

	Bivoltine	Multivoltine
Eggs/dfl	400-450	350-450
Hatching (%)	90	90
Cocoon weight (g)	1.5	1.0
Shell ratio (%)	18	13
Pupation rate (%)	90	90
Number of cocoons/kg	700	1000

To supply disease free eggs, female moths and pupae are examined by mother moth testing/pupal testing for the presence of pebrine spores. The contaminated lots are immediately disposed by burning. Seed cocoons are examined for pebrine infection by the following methods.

Midgut Examination for Pebrine Disease

Protozoan (pebrine) disease is transmitted from generation to generation by female moths. Spores are passed to larvae through contaminated eggs. Spores could enter in the yolk during egg development.

— Collect female pupae and make puncture on the ventral side below the wings
— Press the body gently to squeeze the midgut
— Crush it with 5% potassium carbonate (50 g/1)
— Examine the smear under microscope 600X for pebrine spores

Seed Cocoon Preservation

For preparation of cross breed laying (Fl hybrid), multivoltine races are used as female parent and bivoltine races are used as male parent. Male and female pupae can be sexed by observing the markings in the abdominal area. If this procedure is adopted, we have to cut open all the cocoons to observe the sexual difference. This may lead to wastage of cocoons (male multivoltine) and we cannot use it for reeling. Instead, a simple technique is followed to sex the pupae of bivoltine based on the quantitative character.

— Collect 20 bivoltine pupae and record the individual cocoon weight in a sample
— Work out the average pupal weight
— The cocoons are later selected as female if the weight of the cocoon exceeds average values and the male cocoons with lesser value.
— Female cocoons can be used for reeling

Process of Egg Production

Visit a grainage and note down the different rooms located in the grainage.

— Interact with the grainage inspector and record the purpose of each room.

— Draw a diagram of grainage indicating the place of cocoon storage room, egg laying room, mother moth testing room, cold storage, acid treatment room, oviposition room, incubation room, pierced cocoon room and sales counter.

Selection of Cocoon

Selection seed cocoon with high pupation rate (>90%). Dissect out 20 pupae per batch and examine the gut for pebrine, avoid melted cocoons.

Moth Emergence

— Moths mostly emerge during early hours

— Male moth emerges earlier than female

— Store the male moths in refrigerator for synchronisation in pairing

Multivoltine : 10 days

Bivoltine : 12 days

— To avoid selfing in multivoltine race, pick up the male moths and destroy them. Observe the sexes based on morphology and behaviour.

Female Moths	Male Moths
Bulky with heavy abdomen	Narrow and pointed abdomen
2 paded ovipositor at the	Trident chitinous male organ
end Sluggish	Active

Sex Linked Chracters:

Female cocoon is yellow, male is white. Male is used for preparing cross breeds.

Pairing

— Allow the female moths for few minutes over the sheet to excrete the meconium

— Spread printed egg sheets of 20 squares on trays of 90 x 60 cm @ 8 sheets/tray

— Place in a dark room at a temperature of 24°C ± 1°C maintain humidity at 80%

— Allow female moths for pairing with males for 5 hr in cellule placed over brown sheet

— Cellule is a funnel shaped plastic cup. It facilitates us to place desired races inside cellule and eggs are deposited in a discrete circular area.

Oviposition

Depairing should be done gently to avoid injury to genital organs After depairing, store the male moths in refrigerator at 5-10°C for second time mating. Discard the male moths after second pairing.

Examination for Pebrine (Mother Moth Testing)

Dry Moth Testing:

— Pick up the female moths at random (20%) from egg sheet of lot

— Collect 30 moths, note the lot number and keep it in hot air oven at 70°C for 6 hr.

— Take the moths and add 90 ml of 0.5% potassium carbonate

— Moths are ground for 2 minutes in a moth crusher and centrifuge at 10,000 rpm and allow them to settle for 2 minutes and filter through absorbent cotton. Centrifuge the contents at 3000 rpm for 3 minutes with the sediment and add 2% KOH

— Examine a drop of solution for the presence of pebrine in a microscope at 600 magnification.

Green Moth Testing:

Imediately after egg laying, muscles from thoracic region of female moth is crushed and examined for pebrine.

Sampling test for pebrine

Number of moths in a lot	Number of moths for examination	Number of pebrine sample for rejection
390	All	1
391-500	390	1
501-600	450	1
601-700	480	1
701-800	565	2
801-900	620	2
1001-2000	755	2
2001-3000	865	3
3001-4000	915	4

Disinfection

— Soak the egg sheet in sodium hypochlorite (140 ml/1) solution for 20 minutes

— Wash it in running water for 5 - 1 0 minutes and allow to dry in shade

— Record the date of egg laying and signature of moth examiner on the egg sheet

Acid Treatment for Bivoltine Hybrid Eggs

Bivoltine hybrid eggs undergo hibernation within 48 hr

after laying. To avoid the hibernation in bivoltine hybrids, acid treatment is followed. Two types of acid treatment are given (hot or cold acid treatment). The duration of egg treatment varies with method of treatment.

Hot Acid Method

Prepare HCl at a specific gravity of 1.075 at 46°C for five minutes, Treat 15-24 hr old eggs in acid treatment bath for five minutes.

Cold Acid Method

Treat the egg sheet in HCl (1.110) at room temperature 24°C - 26°C for 30 minutes.

Loose Egg Preparation

This method ensures higher hatching percentage by eliminating sterile eggs.

Preparation

- Coat the craft paper of 90 x 60 cm with starch solution (30 g/1)
- Allow female moths to lay eggs on sheet
- Place the tray in dark room for 24 hr
- Examine for pebrine spores
- Wash the eggs in running water for 5 minutes
- Treat eggs in bleaching powder solution (5 g/1) to remove starch and then with sodium chloride salt solution (Sp. Gravity 1.08).
- Observe the floating eggs and discard them
- Wash the eggs in running water and treat with sodium hypochlorite (140 ml/1) for 20 minutes

— Wash the eggs in running water and shade dry them.

— One g of loose eggs of multivoltine contains 2000 eggs and bivoltine has 1800 eggs.

Post Cocoon Technology

Colour and Shape

Colour of the cocoon is a racial character and it depends on the pigment present in the sericin layer of the silk filament. The colour will be removed along with the sericin layer at the time of degumming.

Shape of Cocoon

It is a racial character. Shape of the cocoon helps to determine the variety and reelability of a cocoon.

Selection of the Cocoon

Good cocoons for reeling are selected based on the compactness, grain, weight of individual cocoon, weight of cocoon shell and shell ratio.

Compactness: When pressed between fingers, it should be firm, compact and slightly elastic Grain : Remove the outer layer of cocoon (Deflossing) and observe for the granules in the cocoon. Coarse granular cocoons make poor reeling.

Weight of the cocoon: Important commercial character and it indicates approximate quantity of raw silk that can be reeled from it. Weight of shell : More important character than the weight of the cocoon.

Larger the weight of the shell, greater is the silk yield.

Shell ratio (SR) : This is calculated based on the formula:

$$\frac{\text{Weight of the cocoon shell}}{\text{Weight of the whole cocoon}} \times 100$$

This value gives a fair indication of the quantity of raw silk that can be reeled from the lot of fresh cocoons. It varies from 12 to 24 per cent.

Identification of Defective Cocoons

For good reeling and to get high raw silk percentage, cocoons should be healthy and free from defectiveness. There are 14 types of defective cocoons. The common types of defective cocoons noticed in Tamil Nadu are:

— *Immature cocoons*: This is due to harvesting of immature cocoons (5 days for cross breeds and 7 days for bivoltine are ideal period for harvesting)

— *Urinated cocoons*: Cocoons will be dotted with black stains. Squeezing of cocoons emanates a bad smell.

— *Loose knit/fragile/ thin cocoons*: The shell is loosely woven with open spaces. Poor in silk content, gets water logged while reeling and become unreelable.

— *Mould cocoon*: Improper storage of stifled cocoon leads to mould attack

— *Double cocoons*: Large, course and irregular cocoons of abnormal shape which have been spun by two or occassionally more than two silkworms. Shells are made up of entangled silk thread and unfit for reeling.

Stifling of Cocoon

Stifling is a process of killing the pupae without, in any way, interferring with the structure of silk shell around it. Steam stifling is the common method adopted. In this method, green cocoons are exposed to the direct action of the very hot wet steam for the required period.

Reeling Water

For cooking and reeling, water plays an important role. It should be colourless, limpid and odourless with pH of 6.8 – 7.4 in cold condition with a total hardness of 35-90 ppm. Reeling unit should be established where there is availability of good quality water. To certain extent, the hardness can be corrected by adding water softening agent (EDTA).

Reeling

The unwinding of silk filament from the cocoons as raw silk. It requires assembling together of the filaments from five or more cocoons depending on the thickness of the raw silk required in the weaving industry.

Cooking

The filaments are held fast by the natural gum sericin. It is necessary to soften the gum by putting the cocoon in hot water before unwinding in the reeling machine. Softening is called as cooking or boiling.

There are two systems of cooking viz., floating system (top reeling) and sunken system. In the floating system, the cocoons are cooked only to the exent the shell becomes wet and they float in water when cocoons are put into the reeling basin. Cocoons are treated in an open pan and they are hand brushed in the same basins and after getting true ends, the cooked cocoons will be transferred to reeling basins where the temperature is lower.

Brushing

The next step in reeling is brushing. This is done to remove the entangled mass of filament.

Reeling

For reeling of silk filament from cocoons, different reeling devices are used

— Country charkha, improved CSRTI Charka – hand driven

— Cottage basin - hand driven or power driven

— Filature reeling basin - power driven with more mechanism

Charkha

Cooking and reeling take place at high temperature and high speed. Inferior and defective cocoons can be reeled more economically and can be fabricated by village carpenter and blacksmith. Cost of charka silk is lower and more suitable for handlooms both for warp and weft. It is also used for weft in power loom.

Charkha Reeling Device

The cocoons are cooked in the basin, the filaments are collected after brushing and passed through a hole (Tharapatti) in ordinary thread guide device. One thread is crossed with another thread to form chambon type of croissure in order to eliminate water from the thread and agglutination of the filament. The thread is passed through a distributor and reeled on a larger wooden reel. Usually four ends are maintained in this device.

Croissure

Intertwining or crossing of threads by twisting in a series of spirals. It makes the filament compact, round and smooth and firmly cements the filament and squeezes out excess water.

Cottage basin

It consists of a cooking unit and a reel bench to which the reeling basins are attached. Hot water to the reeling basin is supplied through taps drawn from the drum fitted on the cooking unit. Each basin is designed to reel six to ten ends. The filament collected in the basin are passed through a porcelain button to clean the slugs and waste and through an independent croissure (Tavellette type) which is more efficient. Finally the thread is passed through a traverse guide to a small reel. Filature (multiend reeling machine) and automatic reeling machine are under public sector.

Re-reeling

The raw silk obtained from cottage basin, multiend and automatic reeling machine are re-round on to a large reel to have uniform in width, and length and to remove excess moisture. It facilitates diamond crossing to form a skein or a hank after lacing.

Lacing

It consists of passing a thread across the skein in such a way as to divide into 5 equal parts with the purpose of keeping the threads in place to ensure the thread lacing unwound easily.

Skeining

Twisting of raw silk several times. Skeining is done by a skeining machine or a turner.

Book Making

Twisted skeins are packed in bundles of 2 kgs and they are known as books. Books are packed in bales. Indian Silk bale is 20 kgs.

Silk Throwing

A process of actual twisting, winding with or without soaking and doubling without drawing of the continuous raw silk fibre to make it into a yarn suitable for knitting, weaving, etc. The sequence of operation involved in the production of silk fabrics are

- Winding (raw silk on to bobbin)
- Doubling
- Twisting
- Twisting setting
- Rewinding
- Warping
- Weaving

Degumming

Removal of sericin layer from the silk yarn.

Bleaching

Removal of colour from the raw silk. Now you will be knowing that colour of the cocoon has no value in reeling unit. Bleaching helps to add desired colour to the silk.

Dyeing

Adding desired colour to filament. Five types of dyes are used i) direct dyes ii) acid dyes iii) metal complex dyes iv) milling dyes and v) selected reactive dyes.

Finishing

Conventional method of finishing includes calendaring, weighing, scrapping, starch and glue finish, etc. to impart/improve certain characteristics like drape, fall, handle, feel, stiffness, weight, etc. Recently special finishes like anticrease and flame retardant finishes are attempted on silk. Now you have observed different processes in post cocoon technology from cocoon selection to yarn finishing. Answer the following questions.

— Quantity of raw silk production (in the visited reeling unit) per month/annum
— Quantity of cocoons used to get one kg of raw silk (Renditta)
— Cost of production per kg of raw silk/yarn
— Quantity of silk used to produce one silk saree (its length and breadth)
— No. of cocoons needed to produce one silk saree
— Time taken to weave one silk saree
— Quantity of silk jarigai used/saree
— Use of floss obtained at the time of cooking and brushing and its cost/kg.
— By-products obtained from the reeling unit.

Denier:

It is a unit expressed to denote the thickness of silk filament. It is the weight of 9000 m of silk filament expressed as gram. Filament denier decreases from outside to inside gradually.

Layers of Shell:

— Outer layer / floss / Blaze: Non-reelabale waste
— Middle compact shell: Reelable
— Iner layer / Pelade layer: Reling waste

Composition of the Cocoon Shell:

Fibroin : 70-81%

Sericin: 19-20 %

Fat and wax: 0.5-1%

Ash and colouring matter: 1.0 - 1.4 %

Eri Silkworm

Ericulture

The rearing of eri silkworms is known as ericulture. It is the only domesticated non-mulberry variety. Eri fabric is commonly known as poorman's silk.

Taxonomic Position and Racial Character

The wild form of erisilkworm, Philosamia cynthia is also known as Attacus cynthia. It is either bi or trivoltine in nature. The cocoon is brick red in colour. P. ricini was evolved from P. cynthia by the breakage of a chromosome during course of evolution. P. ricini cocoon is white in colour and it fetches higher price. It is a multivoltine domesticiated species. Interbreeding of Ricini with Cynthia often occurs and the mixed culture is referred as Philosamia cynthia ricini.

Morphology of Erisilkworm

Adult

Brown and black with pink border, abdomen wooly white, crescent like marks on all wings.

	Male	Female
Abdomen shape :	Narrow	Bulky
Shape of the antenna :	Bipinnate	Bipinnate
Shape of the wing :	Smaller	Bigger
Activity of the moth :	More active	Less active

Egg

Medium sized and oval in shape with very tough smooth shell. Eggs are deposited in 400 to 500 in clusters. Eggs laid by the first two days are more healthy and used for rearing. Eggs are white but the gum gives them a yellow look.

Food Plants

The food plants for the erisilkworms are castor, papaya, tapioca, ber, Ailanthus, Heteropanax fragans (Kesseru plant) etc. Among the food plants, castor is much preferred due to higher water content, ash, nitrogen percentage, acidity and crude protein.

Larval Stage

Eri silkworm undergoes four moultings with five instars. The domesticated S. ricini exhibits several larval strains viz., plain, spotted, semizebra and zebra white. Blue and green based larval body colour also occurs. These strains cross breed naturally. There are several special pigments on prothorax and clasper. The body is with many tubercles and spots. Instars can be fixed based on the marking in prothorax and anal clasper.

First Instar

Length about 5 mm. The head is black and shiny. Rear the first instar from brushing and note down the larval duration.

Second Stage :

Length about 12-15 mm, head black and shiny.

Third Stage:

18-21 mm long. The head is black.

Fourth Stage

It is about 40 mm. The colour of the larvae tends to change in this instar. The body is covered with a white powdery substance or bloom.

Fifth Stage

Larvae eat enormously and grow quickly and can attain 95 to 100 mm long and 15 mm thickness. The general colour of the larvae remain white. Some becomes green.

Cocoon

Eri silk cocoons exhibit colour dimorphism as brick red or white in colour. White cocoons are preferred. Matured larvae leaves an exit hole at one end which facilitates easy emergence of the moth. Eri larvae spin discontinuous filament and moths can be allowed to emerge before reeling. The eri pupae are not killed as done in mulberry silkworm and hence it is known as 'ahimsa silk'. Buddhist monks use only silk fabric made out of eri silkworm.

Rearing of Eri worms

Incubation

One disease free laying contains 300 eggs. Eggs laid by the female on the first two days are selected for rearing. The eggs are treated with formalin 2% on fourth day. Keep the eggs in the moist and cool place. Keep the eggs in bamboo trays covered with wax paper. Incubation at 22°C with high humidity favours higher percentage of hatching.

Brushing

Tiny worms tend to remain together and do not easily

move out. Provide tender leaves. The worms cling on to the under surface. Transfer the leaves to bamboo trays.

Feeding

Provide older leaves to the grown up worms. The leaves should be dry and dust free. Avoid yellow leaves. For the first instar, give five feedings per day (this includes night feeding also). From fourth stage onwards, give three day feedings and one night feeding.

Moulting Care

Follow all the care recommended for mulberry silkworm rearing. A slight lowering of temperature in the beginning and little increase in humidity towards the end of moulting is recommended.

Spacing

Proper spacing helps in better growth and avoids disease epizootic. In the first stage, development of the worm is most rapid. Provide spacing second day after hatching. Adjust the operation along with bed cleaning. Worms require three times the space occupied by its body.

Age	Space covered by the worms	Spacing required for 10 layings (3000 worms) in sq.	
		Beginning	End of the stage
First	4.5 times	1/4	3.5
Second	3.5 times	3.5	7
Third	3 times	7	16
Fourth	3 times	6	40
Fifth	2.5 times	40	80

Bed Cleaning

During first instar, one cleaning is sufficient because of smaller size of the larvae and tender leaves. In the second stage, first cleaning is done after four feedings and second cleaning is done when the worms loose their appetite and prepare to go to moult. In third and fourth stages, first cleaning is done after three feedings in the first day. Second cleaning should be given at maximum appetite period. Third cleaning is given when larvae preparing to go to moult. In the fifth instar, bed cleaning is carried out once in the morning.

Three methods of cleaning are carried out.

— Cleaning with half burnt paddy husk or only husk
— Net cleaning
— Husk and net method.

Paddy Husk Method

A thin layer of husk is sprinkled over the bed. The worms rise above through the husk. It is used when the bed is damp in moist season.

Net Cleaning

This method is similar to that of mulberry silkworm net cleaning.

Husk and Net Method

Combination of both. Worms are lifted after two feedings and transferred to another tray.

Mounting

Transfer the mature worms to bamboo baskets loosely filled with dry paddy straw. Worms spin cocoons in between the folds. Provide proper ventilation.

Harvesting

Worms complete spinning on third day, collect the cocoons in fifth day.

Spinning

Eri cocoons are open mouthed and not composed of continuous filament.They are not reelable. Cocoons are boiled in soda of potash for degumming. After degumming, cocoons are put in bags and boiled in water. Later cocoons are washed in cold water for several times and shade dried. Cocoons thus prepared can be spun in wet condition with takli and in semidry condition with hand charkha (spinning wheel).

Tasar Culture

Tasar silkworm is wild and reared outdoor. Tropical and temperate tasar silkworm are reared in India. Antheraea mylitta is tropical tasar whereas A. pernyi, A. proylei and A. yamamai .are temperate tasar species. A. mylitta and A. proylei species are native to India. Tasar worms are uni/bi/trivoltine. It undergoes pupal diapause.

Food Plants

Tropical tasar silkworm is polyphagous in nature and feeds on Terminalia tomentosa, T.arjuna, Shorea robusta and Zizyphus jujuba.

Taxonomy

Tropical tasar, Antheraea mylitta, Saturniidae, Lepidoptera.

Morphology

Egg is oval and exhibit bilateral symmetry along the anteroposterior axis.

Larva : The larva is tetramoulter. The larvae exhibit different markings in different stages. These spots are well differentiated in prothoracic region and in clasper.

First instar larvae have a black mid dorsal line extending from first to seventh abdominal segments dumb bell shaped black mark on the eight segments. Third instar larvae have a yellow lateral line from second to the tenth of abdominal segments.

Tubercles

The larva has five types of tubercles viz., dorsal, upper lateral, lateral, lower lateral and caudal. They are black in first instar, orange red in second, violet in the third and fourth instars. Number of tubercles present in the larval body differs with first thoracic, second and third thoracic segments and 1st to 7th abdominal segments.

Shining Spots

Silvery white lateral shining spots (oval/triangular) appear on the lateral side from third instar onwards. Some species are without spots. They may have symmetrical or assymetrical pattern.

Cocoon

The cocoon is single-shelled, pendant, oval closed and reelable with non-flossy shell with fine grain. At the anterior end, there is well formed dark brown peduncle with a ring. Cocoons are brown yellow or grey. Larvae spin continuous filament like mulberry silkworm. Each cocoon weighs about 10 - 12 g and measures 6 x 4 cm. The length of filament ranges from 600-1200 m.

Pupa

Cut open the cocoon and observe the obtect pupa. On 8th and 9th abdominal segments, sexual markings are seen female with linear marking and male with round genital aperture.

Adult

Moths exhibit sexual dimorphism. Female moths are bigger with a distended abdomen and narrow bipectinate antennae. The females are polymorphic in colour, grey and yellow. Males are small and brick red or brown coloured. It is a non-feeding stage with vestigeal mouth parts.

Rearing

T. arjuna plantation planted at 1.2 x 1.2 m spacing becomes suitable for rearing within 3 years. Keep the plantation at 2-3 m tall by pruning. Terminalia plantation should be kept clean and base of the each ree trunk covered with any insecticidal dust to prevent the dispersal of the ants from soil to tree. Remove red ant colonies (Oecophylla) in the tree to avoid predation on young larvae.

Brushing

Newly hatched larvae can be directly transferred to tree branches. A small twig should be placed over each newly hatched larva and then tied on the tree for uniform distribution.

The first instar can be reared under controlled condition on cut twigs, kept in a cage. The bottom of the twigs should be immersed in water kept in a bottle to prevent quick drying of leaves. The rearing of larvae can be taken upto 3 instars under nylon netting.

Fourth instars can be transferred to forest plantation. Moults of I to IV instars occur in 3-4, 5-7, 7-8 and 8-10 days respectively while the V instar takes 15 days of voracious feding to become full grown when it measures 12-15 cm and weighs 45-50 g. The cocoons are collected by tribals and stifled.

Stifling and Reling of Cocoons

Tasar silkworm cocoons are hard and hence they are first soaked in 5 % Na_2Co_3 solution for for 18 hours and then subjected to steam cooking in pressure chamber for 21/2 hours. After 24 hours the cocoons are washed in 0.5 % formalin for 15 - 20 minutes. These steps give silk fibres a greater tensile strength. Cocoons are then squeezed to expel water and reled on a reeling machine.

Economics of Silkworm Repairing

Leaf Harvest Method

S.No.	Operation	Cost in Rs./ha
A.	*Establishment of mulberry garden*	
1.	Tractor ploughing 15 h @ Rs.140/h	2,100
2.	Farm yard manure 20 T @ Rs.250/ton	5,000
3.	Land levelling, ridges and furrows 10 pairs @ Rs.100 per pair	1,000
4.	Spreading Farm yard manure 10 men and 10 women	800
5.	Forming irrigation channel and ridges rectification	500
6.	Nursery preparation 10 man days	1,000
7.	Transplanting	1,000
8.	Hoeing/weeding 3.4 times 25 women/time	3,000
		14,400
B.	*First year maintenance*	
1.	Cost of fertilizer 300: 120: 120 NPK kg/ha	5,500
2.	Fertilizer application cost	500
3.	Irrigation 60 men days	3,000
4.	Leaf harvest (30 women)	900
5.	Land revenue	100
		10,000

Contd....

C.	*Second year maintenance*	
1.	Farm yard manure 20 ton/ha @ Rs.250/ton	5,000
2.	Cost of fertilizer NPK 300: 120: 120 kg/ha	5,500
3.	Fertilizer application cost	500
4.	Irrigation 60 men days	3,000
5.	Hoeing and weeding 3-4 times, 25 women/time	3,000
6.	Leaf harvest 500 women	15,000
7.	Pruning 20 women	1,000
8.	Land revenue	100
		33,100

D. Cost of silkworm rearing

		Unit cost	Utility period years	Depreciation value/year
1.	Rearing room	40,000	25	4,000
2.	Rearing stand (15)	8,400	10	840
3.	Chawki rearing stand (2)	500	10	50
4.	Feeding stand (4)	300	10	30
5.	Wooden rearing trays (20)	2,000	10	200
6.	Bamboo trays 160 No. @ 40 / Unit	6,400	3	192
7.	Leaf choping board 2 Nos. knives (2 nos.) leaf chamber (2 nos.)	1,000	10	100
8.	Antwell 60 No.@ 10	60	10	6
9.	Bamboo mountage (100 Nos.) @ Rs.100	10,000	3	300
10.	Foam pads (40 Nos.)	250	5	10
11.	Bed cleaning nylon net 400 Nos.	1,200	3	36
12.	Thermometer 1	10	10	--
13.	Hygrometer 1	500	10	--
	Total depreciation cost/year	71,710	--	5,764

Contd....

E. *Rearing expenses*		
1.	Disease free laying (DFLs) purchase 2500 DFLs @ Rs.2.50/DFL	6,250
2.	Labour for feeding silkworm 450 men	22,500
3.	Cost of paraffin paper, bleaching powder old news paper	1,000
4.	Non recurring depreciation value	5,764
	Total	35,514

F.	*Economics of Sericulture*	
1.	Mulberry leaf yield	35,000 kgs
2.	Cost of leaf production (A+B+C)	57,500
3.	Cost per kg of leaves	1.65
4.	Sales of cocoon 1500 kg (60 kg/100 DFLs) @ Rs.100/kg	1,65,000
5.	Cost of cocoon production	35,514
6.	By product value	2,500
7.	Total expenditure of silkworm rearing (A+B+C+E)	93,014
8.	Net income (4-7)	71,986
9.	Cost of production of 1 kg of cocoon	62

Shoot Rearing Method

A. *Establishment of mulberry garden:*

Sl. No.	Operation	Cost in Rs./ha
1.	Tractor ploughing 15h @ Rs.140/h	2100
2.	FYM 20 T/h @ Rs.250/ton	5000
3.	Land levelling, ridges and furrows 10 pairs @ Rs.100 per pair	1000
4.	Spreading FYM 10 Men and 10 Women	800
5.	Forming irrigation channel and ridges rectification	500
6.	Nursery preparation	1000
7.	Hoeing/weeding 3-4 times 25 women/time	3000
		14400

9

Quality of Reeling Water

Approximately 850-1,000 tons of water is used to manufacture 1 ton of raw silk. All natural water sources contain minerals. The mineral constituents found in greatest abundance are bicarbonates, sulphates, calcium chlorides, magnesium and sodium. Carbon dioxide contents may be greater than expected as a result of the decomposition of organic matter. Three water sources are described here: well/spring water, lake water and surface water/river water.

a) Well water is usually clear with a consistent composition. It may contain sodium bicarbonate, calcium bicarbonate, magnesium bicarbonate, iron and carbon dioxide.

b) Lake water can be clear but may be coloured and slightly acidic. High acidity combined with the presence of dissolved gases makes the source corrosive. Although lake water contains some calcium, magnesium chlorides and sulphates, the concentration of electrolytes is low.

c) Surface water contains sulphates, chlorides, calcium bicarbonate and magnesium bicarbonate. The content of minerals and impurities is determined by the profile of the area from which the water originates, plus seasonal changes. Water quality is affected by the amount of industrial affluent discharged into canals and rivers.

Impurities in Water

Impurities in water are discussed under the following headings, a. Turbidity and colour b. Alkalinity c. Hardness d. Iron and manganese

a) *Turbidity and colour.* Turbidity may be caused by large or small mineral and organic particles suspended in the water. Mineral particles may include clay, silt, calcium carbonate and silica. Organic particles may include fine vegetable wastes, fats and microorganisms. Sedimentation in tanks or reservoirs is adequate to clear large particles, but filtration is needed for small or colloidal fragments. Filtration using sand or the use of coagulants followed by filtration may be indicated. The colour in water signals the presence of dissolved or colloidally dispersed organic matter of unknown composition augmented colloidal iron or manganese.

b) *Alkalinity.* Raw water contains bicarbonates in amounts dependent on their origin. Often, water may contain small amounts of carbonate alkalines. The number of bicarbonate and carbonate ions present may be isolated by titration using phenolphthalein as an indicator and methyl orange as an indicator in the second stage.

c) *Hardness of water.* Water hardness is caused by calcium and magnesium salts. Other contributing elements to hardness include metallic ions such as iron and strontium. Note that if metallic ions are present they will be found in minute quantities compared to salts.

Hardness is described as permanent and temporary; where permanent is caused by nitrates, chlorides and sulphates, and bicarbonates, which may be boiled to precipitate into carbonates and removing the hardness, cause temporary.

$$Ca(NCO_3)_2 \rightarrow CaCO_3 + H_2O + CO_2$$

Hardness can be measured when a given amount of potassium palmitate or oleate solution is titrated using a given amount of water, concluding when the shaken liquid produces a permanent lather.

More sophisticated methods are now available where the concentrations of calcium and magnesium ions are quantified by titration with ethylene diaminetetra-acetic acid. Hardness is expressed in degrees, but the actual units of measure vary from country to country.

1 unit ($CaCO_3$ ppm) of United States is equivalent to 0.056° dH of Germany.

d) *Iron and manganese.* Iron and manganese may be found in water depending on the source of the contamination. The three types of impurities are: (a) ferrous and manganous bicarbonates in well waters when high amounts of free CO_2 are present, (b) ferrous and manganous sulphates in rivers containing acid mine-waste waters, or (c) iron and manganese mixed with organic waste. Iron is present in some waters at the source, but corrosion in pipelines and storage vessels may taint water.

If a given source is alkaline, iron in the form of ferrous bicarbonate may be removed by aeration when ferric hydroxide is precipitated and carbon dioxide released. Aeration also helps to raise the pH by reducing the content of dissolved carbon dioxide.

$$4Fe(HCO_3)_2 + O_2 + 2H_2O \rightarrow 4Fe(OH)_3 + 8CO_2$$

Manganese can be released in the same way but to ensure rapid oxidation, a pH greater than 10 is required through the addition of lime or caustic soda.

Methods of Water Softening

This section looks at industrial water treatment using lime-soda processing, base exchange and demineralisation to remove all dissolved salts.

a) *Lime-soda process*. In this method, hydrated lime and sodium carbonate are added to precipitate calcium and magnesium ions as compounds of low solubility. For temporary hardness arising from bicarbonates, the reactions are as follows:

$$Ca(HCO_3)_2 + Ca(OH)_2 \rightarrow 2CaCO_3 + 2H_2O$$

$$Mg(HCO_3)_2 + 2Ca(OH)_2 \rightarrow 2CaCO_3 + Mg(OH)_2 + 2H_2O$$

Calcium carbonate has a much lower solubility than magnesium carbonate. When water containing bicarbonates of calcium and magnesium is treated, it is the calcium carbonate, which is first deposited. For permanent hardness, the reaction is:

$$CaSO_4 + Na_2CO_3 \rightarrow CACO_3 + Na_2SO_4$$

$$MgC1_2 + Ca(OH)_2 \rightarrow CaC1_2 + Mg(OH)_2$$

The $CaC1_2$ formed in the last reaction is removed as carbonate by the sodium carbonate. It may be noted that, when temporary hardness is eliminated by this method, no salts remain but permanent hardness, an equivalent quantity of sodium salts is left behind.

b) *Base or cation exchange.* This process relies on the exchange of calcium and magnesium ions, which occur when water is passed through a bed composed of complex aluminium silicates commonly known as Zeolites.

If Z represents the silicate, the reaction may be represented as follows:

Temporary hardness

$$Ca(HCO_3)_2 + Na_2Z \rightarrow CaZ + 2NaHCO_3$$

$$Mg(HCO_3)_2 + Na2Z \rightarrow MgZ + 2NaHCO_3$$

Permanent hardness:

$$CaSO_4 + Na_2Z \rightarrow CaZ + Na_2SO_4$$

$$MgSO_4 + Na_2Z \rightarrow MgZ + Na_2SO_4$$

In contrast to the lime-soda process, the exchange process replaces the calcium and magnesium ions by an equivalent of sodium even when the hardness is temporary. The anions present are unaffected. The bed is made from Zeolite in granular form, sieved to a particle size around 1-2 mm in diameter; its capacity is limited and will eventually be depleted.

However, the process is reversible and employing a concentrated salt solution can regenerate the bed:

$$CaZ + 2NaC1 \rightarrow CaC1_2 + Na_2Z$$

When the calcium and magnesium ions are removed as soluble chlorides, they are replaced by sodium. A cyclic procedure is created of softening, regeneration, water wash (to remove excess salt), softening, etc.

Between the softening and regeneration, it is sometimes desirable to backwash by passing water upwards through the bed for a short time to remove any matter, which has filtered out of the bed. Water from this process may have hardness values as low as 0.5 ppm.

c) *Demineralisation.* If two exchange processes are used, one for the anions and one for the cations, the content of mineral salts may be reduced to that present in distilled water. Removal of the metal ions may be carried out with one of the sulphonated resins in the "Hydrogen" form, so that water running out contains hydrogen instead of metal ions.

$$CaC1_2 + H_2R \rightarrow CaR + 2HCL$$

$$Ca(HCO_3)_2 + H_2R \rightarrow CaR + 2CO_2 + 2H_2O$$

The resultant carbonic acid breaks down into water ad carbon dioxide, which can be removed by aeration, but the mineral acids remain. Regeneration of the exchange process may be carried out by the use of hydrochloric acid.

Table 1. Ion-exchange types and the quality of water treated

Type of treatment	*Original water*	*Base type*	*Hydrogen type*
pH	6.4	6.8	3.3
M-alkalinity (CaCO3 mg/1)	60	81	0
Acidity (CaCO3 mg/1)	30	0	74
Whole hardness (CaCO3 mg/1)	88	8	2

Standard Quality of Reeling Water

It is essential to carefully select the quality of reeling

water, as it impacts reeling efficiency, raw silk percentage of cocoons and the raw silk quality. Table 3 gives a standard for quality of reeling water and Table 3 shows the pH values and water hardness desirable for cocoon cooking water.

Table 2. Standard quality of reeling water

Items	*Standard concentration*	*Range of concentration*
(1) Colour and cleanness	Colourless and clear	
(2) smell	No smell	
(3) Suspension and sediment	No	
(4) pH of water	6.9	6.6-7.2
(5) pH of water after being boiled	8.3	7.9-8.6
(6) specific electro conductivity (micrombo/cm)	100	40-300
(7) Hardness (°dH)° dH x 17.85 - $CaCO_3$ ppm	2.0	0.5-4.0
(8) M-alkalinity ($CaCO_3$ ppm)	30	20-40
(9) Total acidity ($CaCO_3$ ppm)	5	3-15
(10) Heavy metal iron (Fe_2O_3 ppm)	Under 0.1	0-0.3
(11) Residue after evaporation (ppm)	85	50-200

Analysis of Reeling Water

a) pH. A pH meter can measure the value of pH.

b) *Acidity.* A 50 ml water sample is taken from the original water and 5 drops of phenolphthalein indicator added. Titrate with the N/50 of NaCO standard solution until the yellowish colour of the sample water changes to purple.

Table 3. pH values and water hardness suitable for cooking parts

Parts	*Items*	*Good reelable cocoons*	*Poor reelable cocoons*
Dipping part	pH	4.5-5.5	4.5-5.5
Water hardness	(°dH)	0	0
Low temperature permeating part	pH	5.5-6.5	6.0-7.0
Water hardness	(° dH)	0-2	0-2
Cooking adjusting part	pH	5.5-6.5	6.5-7.5
Water hardness	(° dH)	2-4	1-3
Finishing part	pH	6.5-7.0	7.0-8.0
Water hardness	(° dH)	1-3	1-2

$$a = b \times \frac{1,000}{e}$$

a, Total acidity ($CaCO_3$)

b, Amount of N/50 NaOH sol. Consumed for titration (ml)

c, Amount of sample water (ml)

c) *M-alkalinity:* A 50 ml of water sample is taken from the original water and 5 drops of methyl red indicator added. Begin titration by using the N/50 N_2SO_4 standard solution. Terminate when the bluish colour of the water sample changes to purple (pH 4.8).

$$a = b \times \frac{1,000}{c}$$

a. M-alkalinity ($CaCO_3$)

b. Amount of N/50 H_2SO_4 sol.Consumed for titration (ml)

c. Amount of sample water (ml)

d) *Water hardness:* A 50 ml water sample is poured into a beaker. pH 10 is reached by adding 1 ml of Buffer solution and a few drops of E.B.T. indicator to the water sample, shaking it for a while and then begin the titration by using the E.D.T.A. standard solution. Terminate when the reddish colour of the water sample changes to a bluish colour.

$$a = b \times \frac{1,000}{c} \quad {}^\circ dH = a \times 0.056$$

a. Total hardness ($CACO_3$)

b. Amount of N/50 H_2SO_4 sol. Consumed for titration (ml)

c. Amount of sample water (ml).

10

Cocoon Cooking

Cocoon cooking unwinds the cocoon filament spun by the silkworm. The sericin covering around the cocoon filament is agglutinated after silkworm spinning, then hardened through the cocoon drying process. In preparation for reeling, it should be softened. Processing softens sericin by heat, water and steam. Ideally there will be uniform softening of the outer and inner cocoon shell.

Cocoon Cooking Methods

Pan Cooking

a) Cocoons are put into a pan of boiling water. The groping ends of the cooked cocoon can be completed with the stirring rod in the pan.

b) The cocoons contained in the wire cage are placed into the boiling water and then boiled for a few minutes. After boiling, the cocoons with the wire cage are moved into a low temperature bath in another pan. If there is water permeation inside cocoon shell and swelling of the cocoon shell, this work can be repeated.

Machine cooking

Recently, machine cooking has become widely used in most silk reeling factories (Figure 3). In general, the machine cooking process is divided into six parts as follows:

Soaking part

The surface layer of the cocoon swells over at the soaking part. The first part is carried out by dipping the outer layer of the cocoons into a water bath at 55°C. This step is comparable to fabric dipping for dyeing to ensure even results.

High temperature and low temperature permeation part

The uniform cooking of the outer and inner layers of the cocoon can be easily attained by replacing the air of the cocoon cavity with water since water transfers heat faster than air.

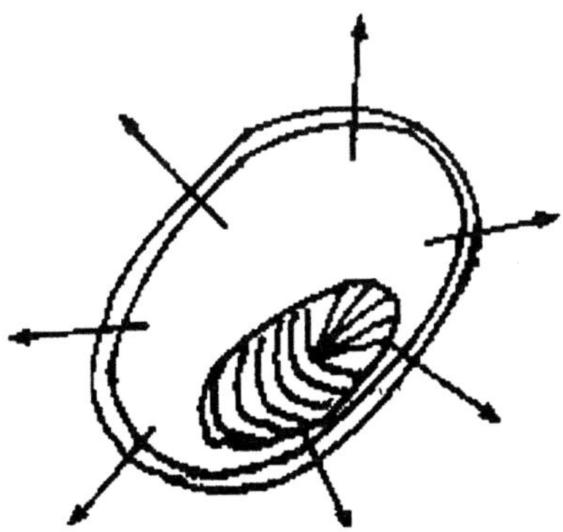

Figure 1. Air expansion in cocoon cavity and partially replaced with steam at high temperature permeation part.

Thus, the wet cocoons from the soaking part are exposed to steam at about 90-95°C at an appropriate steam pressure. Next, the air inside the cocoon cavity is heated. When cocoons treated at high temperature permeation are moved into low temperature permeation around 65°C, partial condensation occurs in the cocoon cavity. Then the cocoon sucks water in, evenly wetting all the layers of the shell.

The amount of water permeated into the cocoon cavity is controlled by the difference in temperature between high temperature and low temperature permeation parts, as well as the air permeability of the cocoon shell.

Steam cooking part

Now the cocoons treated in steps a) and b) are moved to the steam processing. This part causes the sericin to swell and soften the silk layers and the steam to fill up the cocoon cavity by diffusing the permeated water out of the cocoon. Steam heat is a highly efficient head conductor that can cause unwanted sericin loss. To minimise or stop this loss of sericin, infra-read rays have been attempted in this process, but proven unfeasible. For cocoons anticipated having poor reelability, the steam cooking part has to be prolonged.

Sudden variations in steam pressure can adversely affect the cooking process by producing over processed or insufficiently processed cocoons. These poorly cooked cocoons seriously deteriorate reeling efficiency by decreasing raw silk yield and quality and boosting cleanness defects during reeling. It is urged that greater vigilance be exercised to control steam cooking.

Cooking adjustment part

At this point the steam content of the cocoon cavity is

replaced with water through gradual condensation of steam in the cocoon. This is effected by gradually cooling of the water from 98° to 65°C.

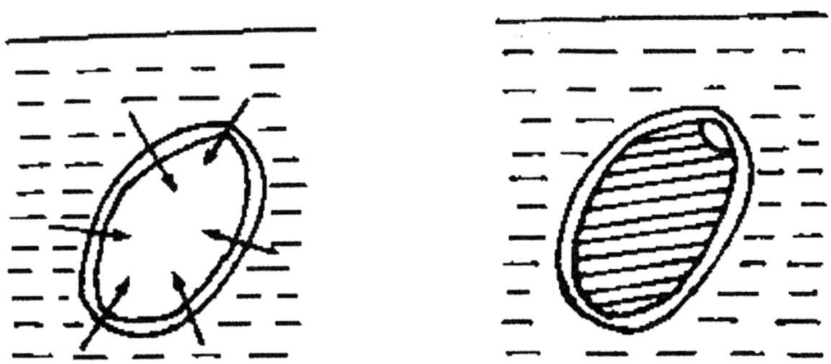

Figure 2. Cocoon sucks in water at low temperature permeation part

Sericin swollen by steam cooking becomes stable. As this step consumes large volumes of water for cocoon permeation, it needs to be longer in duration and requires more fresh water than other parts.

Low temperature-finishing part

Cocoons are finished in 50-60°C water. Here, the cocoons, which were properly swollen by processing in the first five steps, become more stable and are prepared for the next stage in reeling.

Degree of Cocoon Cooking

Optimum degree of cocoon cooking can be gauged correctly by the reeling results, but it also can be evaluated as follows:

a) *Response to touch:* In the case of optimum cooking, the cooked cocoon is velvety and the shell is neither soft nor hard when felt by the hand. In the case of over cooking, the cocoon is soft and

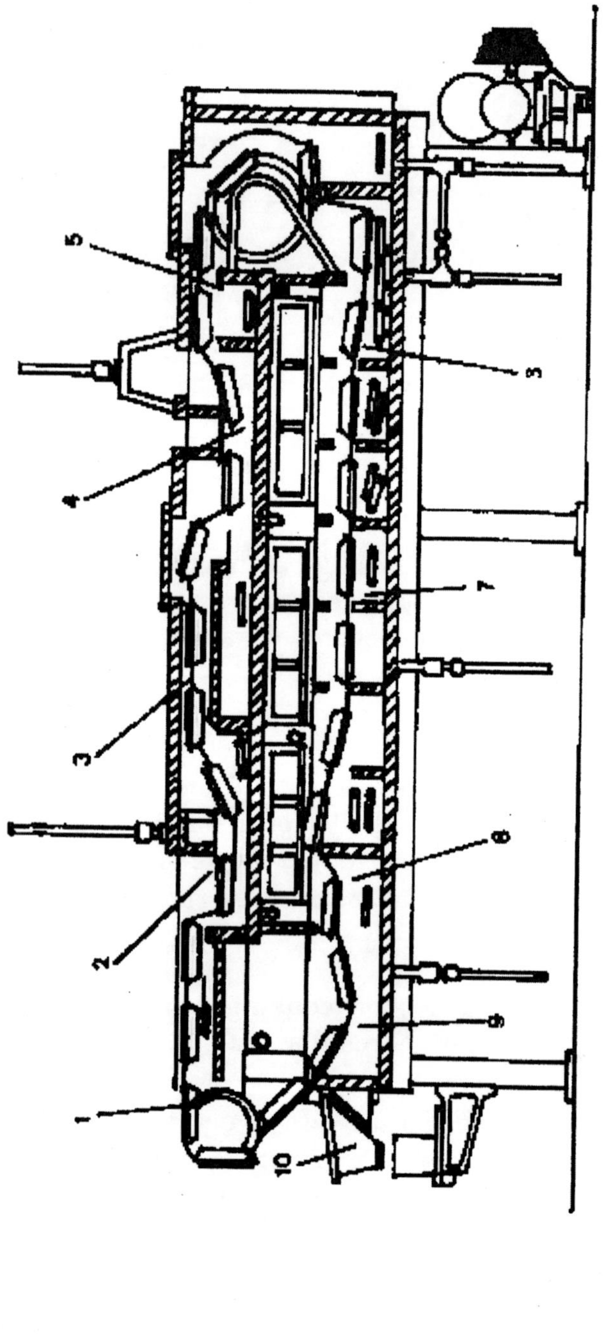

Figure 3. Cocoon Cooking Machine1.Cocoon hopper part, 2. Soaking part, 3. High temperature permeating part, 4. low temperature permeating part, 5. Cooking part, 6-7. Adjusting part, 8-9. Low temperature finishing part, 10. Cocoon outlet

collapses easily under finger pressure. If under or imperfectly cooked, it is hard.

b) *Groping and condition:* In case if optimum cooking, the filament unwinds by itself at the moment the groped end is picked up and grasped by hand, with little resistance on the fingertip, and without breaking the end.

Table 1. Criteria of insufficient and overcooking

Items	*Insufficient*	*Overcooking*
Colour and touch of cooked cocoons	Creamy and rough	Grey-yellowish and soft
Conditions of groping ends	Hard	Not difficult
Number of dropping ends in reeling process	Increase	Decrease
Increase of by-product silk	Parchment layer	Brushing waste
Kinds of commonly occurring defects in raw silk	Split ends and loop	Slug
Degumming ratio of raw silkHigh	Low	
Tangle of cooked cocoons	Slight	Heavy

c) *Cocoon colour:* In imperfect cooking, it is creamy and cooked partially. When over cooked, the cocoon is grey-yellow.

d) *Cooked cocoon shell weight:* five to six times as heavy as the original shell weight is desirable; 4.5 times or less is defined as under or imperfect cooking, while the overcooked cocoon weighs six times or more than the original.

e) *Cooked cocoon weight:* 10-11 times the weight of the original dried cocoon.

f) *Conditions of cooked cocoons in basket:* The cooked cocoons should easily drop out of the basket, when the baskets are opened at the end of the cooking process.

g) *Crushed cocoon:* There are no crushed cocoons in optimum cooking.

Adjustments to Cooking Conditions

Cooking methods should be re-adjusted for better results depending on the different quality of cocoons as shown in Table 2.

Table 2. Adjustment of cooking methods depending on variations in cocoon quality

Different cocoon qualities	*Re-adjustment of cooking methods*
Thick shell cocoon with high reelability	In pre-treatment (soaking, steaming, permeation) difference of temperature should be more, while steam pressure and temperature of steam cooking part should be lower
Thin shell cocoon with high reelability	Different of temperature part should be higher and steam pressure and temperature of steam cooking part lower
Thick shell cocoon with poor reelability	Difference of temperature in pretreatment as well as steam pressure and temperature of steam cooking part should be more.
Insufficiently dried cocoons	Steam pressure and temperature of steam cooking part less
Over dried cocoons	Steam pressure and temperature of steam cooking part more

During treatment, the temperature variations should be less for good reelable cocoons than for poor reelable ones, where steaming pressure and temperature of steam cooking should be increased. The temperature in cooking adjustment should be precisely maintained for good reelability cocoons. This level of precision is unnecessary for cocoons with poor reelability. Remedies for poor cooking results are shown in Table 3.

Table 3. Points for cooking operations

Phenomena	*Counter measures*
Crushed cocoons	Reduce the difference of temperature between high and low temperature permeation parts as well as steam cooking part and first cooking adjustment
Overcooking	Lower the temperature of steam cooking and high temperature permeation parts Shorten the time for cooking Modify the quality of reeling water
Insufficient cooking	Increase gradually the temperature of steam cooking part Prolong the time for cooking Use the soluble type water for cooking instead of hard type one
Floating cocoons	Make the temperature curve in cooking adjustment Increase the difference eof temperature between high and low temperature permeation parts Reduce the temperature of low temperature finishing part
Uneven cooking	Increase the temperature of soaking part Check the steaming pipe in the high temperature permeation part Strict assorting of inferior cocoons

Reeling Result

The suitability of cocoon cooking has an important

bearing on the reeling result. A high proportion of waste is caused whenever cocoons are over- or undercooked. For overcooked cocoons, this happens when the filament comes out in lumps, clogs the reeling button and increases the number of breaks during reeling.

Table 4. Effects of cooking degrees on reeling results

Degrees of cooking	*Raw silk % of cocoon*	*Reeling troubles per 10,000 m (times)*	*% of by-product silk*		*Neatness (%)*	*Cleanness (%)*
Under cooking	16.00	1.47	1.28	1.23	92	91
Optimum cooking	15.98	1.60	1.45	1.10	94	95
Over cooking	15.94	1.67	1.54	1.03	92	93

Each time a break is repaired, the silk filament must be pulled out and all of the silk removed becomes waste. The undercooked filament does not unravel easily and also causes breakages and waste. In each case reeling efficiency is diminished.

11

Biomass Gasification in Silk Reeling

Biomass was the first household energy source used by human beings. Until the discovery of low-cost petroleum and natural gas in the early 1900s, wood was the most used energy and heat supplier for mankind. In developing countries, like India, biomass is still the main source of energy for village-level cottage industries and institutions. This is largely due to the easy availability and cheaper price of biomass at local levels, compared to other fuel sources. Presently, firewood is collected from the forests and considered to be available for free. The conventional methods employed for burning biomass are also the least efficient. In addition to being less efficient, these practices lead to emissions of smoke, suspended particulate matter, and unburned solid residues; this in turn leads to local-level health problems and environmental degradation.

With the economic development of an area the costs of collection and use of biomass increases, which eventually leads to a switch to fossil fuels. One of the other factors that leads to the switch to fossil fuels from biomass is the connivance of storage and use of fossil fuels, due to their higher energy density. Thus, whoever can afford it, and wherever fossil fuel is available, there is a tendency to shift to it. Gasification of biomass prior to combustion in order to meet the thermal energy needs of the villagelevel cottage industry addresses most of the problems associated with the use of biomass directly. Due to the increase in efficiency, the use of biomass as a source of thermal energy remains viable, even if there is an increase in the cost of collection and use of biomass. It leads to almost complete elimination of smoke and particulate matter generation. It offers the connivance of usage of fuel for energy. Appreciating the benefits which gasification prior to combustion offers, there has been considerable efforts world over towards development of biomass gasification technology and biomass gasifiers. In India the development of biomass gasification has received serious attention with establishment of biomass research centers and gasifier action centers at various locations spread all over the country. These institutions have played a key role in the upgrading and adoption of suitable technologies, including testing, monitoring, and development of biomass gasification systems. However, the focus of these research centers has largely been on large-sized gasifiers, which are suitable for large industries and to fuel biomass-based power plants.

In spite of these efforts not much progress has been achieved towards the commercialisation of largescale biomass gasifiers. One of the reasons often cited for this is that the technologies and gasifiers developed are more generic in nature and do not address application-specific needs. The other reason often cited is that installation of a large-sized biomass gasifier at a location leads to sudden increases in the demand for biomass. This then leads to an increase in the price of biomass, rendering the operation commercially non-viable.

The Energy Resource Institute (TERI) has developed biomass gasification technologies for village-level cottage industries and applications where biomass was already being used, but in a much less efficient way. The technologies developed are industry and application specific, and this strategy has not only helped the local environment but also reduced the health hazards that workers are exposed to. This strategy also ensures long-term, continued use of biomass for energy needs, due to

— improvements in efficiency and consequent reduction in the cost of energy,

— connivance in the usage of energy,

— the use of already established market channels for supply and procurement of the biomass.

The village-level cottage industries use largely fuel wood for thermal heating applications, which is normally collected from the forests or community forests. Wood may also be procured from farm forestry projects. Fuel wood contributes nearly 60 per cent of the total biomass energy in India, but the demand for wood usually outstrips the supply. This leads to unsustainable wood harvesting and degradation of forests.

The objectives of the strategy of developing application and industry-specific biomass gasification technologies, where biomass was already being used are to improve the efficiency of the utilisation of energy, thereby improving the economic viability of industries, leading to sustainable development of rural areas; stop degradation of forests by reducing the consumption of wood; provide a more user-friendly fuel to prevent a shift to fossil fuels; enable consumption of agro-residue in place of wood for given applications; improve the local environment and working conditions for workers.

In India, being an agrarian country, there is easy access to biomass that can be used to generate energy. Burning is the easiest and oldest method used for generating energy and also the least efficient. Firewood is one of the main fuels used by cottage-level industries in India, particularly in the energy-starved rural areas. Traditionally, firewood use in the cottage-level rural industries is characterised by the low efficiency of end-use devices.

Keeping in mind that gasification is probably the most important and efficient energy-conversion technology available for burning a wide variety of biomass fuels, including firewood, TERI developed application-specific gasification systems for a number of rural cottage industries. Biomass fuels, such as firewood and agro-residues, essentially contain carbon, hydrogen, and oxygen, along with some moisture.

Under controlled conditions, characterised by low oxygen supply and high temperatures, most biomass materials can be converted into a gaseous fuel, known as producer gas, which consists of carbon monoxide, hydrogen, carbon dioxide, methane, and nitrogen. This thermochemical conversion of solid biomass into a gaseous fuel is called biomass gasification. The producer gas produced has a low calorific value (1,000 to 1,200 kilocalorie per normal cubic meter (kcal/Nm3)), but can be burnt with a high efficiency and a good degree of control without emitting smoke. Each kilogram of air-dry biomass (10 percent moisture content) yields about 2.5 normal cubic meters (m3) of producer gas. In energy terms, the conversion efficiency of the gasification process is in the range of 70 to 80 percent.

TERI has taken a lead in promoting biomass gasification and developing small biomass gasifiers suited to the requirements of individual industry and applications. Ovens, kilns, and furnaces based on biomass gasifier technologies have been developed for industries, such as silk reeling, textile dyeing, magnesium chloride production, brick drying, and drying spices. Large community cooking systems have also been developed using biomass gasifier technology. These systems have demonstrated fuel wood savings on the order of 50 to 60 percent, along with a significant increase in overall productivity at the enterprise level. The following is a list of the industry and application specific biomass gasifier-based products where thermal applications have been developed.

1. Silk reeling
2. Textile dyeing
3. Large cardamom curing
4. Large-scale cooking
5. Magnesium chloride production
6. Carbon dioxide production
7. Green brick drying
8. Crematorium
9. Rubber drying.

Biomass gasifiers are used in silk reeling to improve the productivity and profitability of the post-cocoon processes in the silk industry. The activities of the project are broadly categorised into

— technology and system development
— economic assessment
— market development
— entrepreneur development
— human resource development.

Under the system development heading, system configurations (packages) of silk reeling and dyeing applications have evolved through a series of prototype development and testing. The commercial model of a gasifier-based silk reeling oven and dyeing system was developed with inputs from silk experts, users, engineering consultants, and manufacturers, and currently commercially marketed by two licensee manufacturers. Both these gasifier-based systems were extensively field-tested to quantify the benefits over the conventional system. An economic assessment of the systems revealed that the system pays for itself in less than a year.

Under the market development heading, several cluster-level studies were carried out to decide the most suitable capacity of gasifier for each application and to make a realistic estimate of the market potential. Commercial partners were selected to cater to the ultimate market demand. A team, consisting of two international consultants and representatives of the project team, surveyed 15 manufacturers to assess their technical, financial, and marketing capabilities. Based on their recommendations, two manufacturers-2M Industries (Mumbai) and Silktex Industries (Kannakpura)-were selected for commercialisation. The technology has been transferred to both the manufacturers through a licensee agreement.

Promotion of Gasification Technology

Towards the development and commercialisation of these technologies work was done in close cooperation with the industries and local-level engineering companies. The strategy of developing application and industry specific biomass gasification technologies, where biomass was already being used, has not only helped the local environment but also reduced the health hazards workers are exposed to.

Promotion of application-specific biomass gasification technologies for small and cottage industries requires considerable market development efforts. The market for the technologies and equipment was created through raising awareness, training, field demonstrations, and capacity building.

The development and promotion of application-specific biomass gasification processes has been done with participation from the local-level process technology experts and industries. Such local-level participation ensures that the system developed addresses the process-specific requirements and is sized appropriately.

For commercial production of the system to be developed, participation from local-level engineering fabricators was invited. Local-level fabricators are in a position to utilize fabrication materials available locally and provide after-sales support. Local government bodies, like the horticulture department, spice board, etc., and funding agencies, like ISPS and the Swiss agency for development and co-operation (SDC), were also major partners.

Impacts of Biomass Gasification

Technology upgrades for meeting the thermal energy requirements of the cottage industry in rural areas using biomass-based energy resources has multiple positive effects. The gasification of biomass prior to combustion improves the economic viability of the cottage industries, thus contributing towards sustainable rural development. Smoke and particulate matter created by the combustion of biomass is eliminated. This not only improves the local environment but also leads to reducing health hazards, thus improving working conditions.

With the improvement in the energy efficiency of these operations, the demand for wood is reduced, thus lowering deforestation levels caused by the use of firewood. The process of gasification enables the use of woody agro-residue and other agro-residue as a source of energy instead of wood, thus further reducing the demand for wood. Extensive performance monitoring of the systems developed thus far under field conditions resulted in fuel wood savings of more than 60 percent. The practice of gasification also results in improvement in the quality of the product and reduction of waste through better control of heating. The improved quality also helps boost the income of rural enterprises.

The present low-efficiency, direct-combustion of biomass for thermal applications in a wide range of cottage industries in rural areas can be replaced with high-efficiency biomass gasification processes, provided the technologies and processes are developed for the specific applications. Such an approach also addresses the issue of suitability of locally available biomass in a particular design of gasifier.

The replacement of existing low-efficiency biomass for energy practices with better technologies, which are more efficient, reduces the requirement of fuel and keeps the use of biomass economically viable-even with increases in the price for biomass. This prevents the shift from biomass fuels to fossil fuels in the process of development of the region.

Improvement in the overall thermal efficiency of the firewood-based heating applications leads to reduction in the demand for wood, which in turn prevents degradation of the forests. The process of gasification of biomass prior to combustion enables the use of agro-residue for the applications where wood was formerly being used. This again prevents further degradation of forests. Gasification of biomass prior to combustion for thermal heating applications, apart from increasing efficiency, also takes care of the emissions of smoke, particulate matter, and unburned biomass, thereby improving the local environment and reducing health hazards workers are exposed to. Some of the drawbacks associated with this strategy of biomass gasification for meeting energy requirements can be addressed by focusing on very small and end-use specific gasification technologies and equipment.

There are some disadvantages associated with the use of biomass as a source of thermal energy. Biomass gasification is by far a cleaner energy production technology, which addresses some of these disadvantages. The development of application-specific technologies demonstrated that the disadvantages associated with the use of biomass for energy applications could be mitigated with both engineering measures and operational "good practices."

Gasification, as an energy production technology, can be regarded as cleaner energy production. It has the uniqueness of being a technology that can even convert waste-from Municipal Solid Waste (MSW) to agricultural or crop residues, like coconut shells, rice husks, straw, wood residues, bagasse, etc.-into a useful and high quality energy source.

The strategy of promoting industry and application-specific biomass gasification technology, after being used initially for the silk reeling industry, has been successfully applied for a number of other applications, like cardamom drying and rubber curing. The application potential of the strategy and the practice has already been demonstrated.

The initial effort, for development of the application-specific "end-use packages," requires funding. This is largely because the present end-users, being largely rural and small entrepreneurs, cannot afford the cost of technology development. The processes and technologies developed for an application, being economically viable, are self-sustainable after initial funding is provided for development, demonstration, training, and commercialisation. End application-specific biomass gasification technologies can be developed for a number of thermal energy requirements in the small- and cottage-level industries in rural areas.

12

Designing of Silk Reeling Factory

The scale of the factory is also influenced by the properties of raw material cocoons because each has different raw silk yieldability. Generally, the number and scale of reeling machines for installation is derived from the daily target of raw silk production, which is calculated by dividing the annual target of raw silk production by 300 working days per year. In this example, the raw silk yield from fresh cocoons is considered for two species.

The raw silk percentage of fresh bivoltine cocoons is applied in Japan and the Republic of Korea, while the Renditta of multivoltine ratio of fresh cocoons required for producing 1 kg of raw silk is used in India, Pakistan and Sri Lanka. In the case of 60,000 kg of annual raw silk production, the required amount of fresh cocoons is as follows:

— Multivoltine cocoon – 540,000 kg

— Bivoltine cocoon- 353,000 kg

The scale of the reeling factory should be differentiated by the raw silk percentage of the cocoon. To summarise, cocoon drying facilities are less useful in tropical countries as fresh cocoon production continues all year round. Temperate Zone countries can only rear silkworms twice or three times a year.

Installation Equipments

The proper equipment for installation has to be selected by considering the following factors:

Properties of Cocoons to be Provided

Length of cocoon filament, length of unbroken filament, size and weight of single cocoon filament, percentage of eliminated cocoons, percentage of cocoon shell and raw silk percentage.

Supply Conditions of Fresh Cocoon

a) Supply times per year of fresh cocoon.

b) Distance, road conditions, vehicle and containers for transportation of fresh cocoons.

c) Quality distribution of fresh cocoons with different crop seasons and production sites.

d) The proper kind and capacity of dryer, adaptable method of cocoon storage, size of cocoon storage room and space for fresh cocoon preparation can be decided based on conditions a to c.

Location and Space

a) Location: The location of the silk reeling factory should be chosen by evaluating availability of reeling water sources, manpower, electricity and transportation. Ideally, the factory will be sited on a slight hill near a good water supply with adequate drainage, near a densely populated region. The main building of the factory is usually in an east/ west direction.

b) Building site. Space for the building site should be procured to minimise the fixed capital and management costs associated with the idle land, unless future expansion is planned at the same location. Most facilities offer sleeping quarters for two shift operations and refreshment for the workers to increase productivity.

Quality and Quantity of Filature Water

As water has a major impact on raw silk yield, it is recommended that the factory have access to well water and surface water sources. Quality and quantity of available reeling water is the most important factor in choosing location.

If a site near well water is considered, pumping trials should be conducted over a two-week period during the dry season. Remember that the amount of reeling water required for production of 1 m/t of raw silk is 1 000 m/t in the Multi-ends reeling machine and 800 m/t in that of the Automatic reeling machine.

Equipment Selection

Drying Machine

a) *Chamber dryer or cabinet dryer.* The delivery and feeding of fresh cocoons to the dryer is carried out manually. Only this type of dryer can execute a small production of fresh cocoons.

— *Hot air dryer*

— *Multi-step type:* The Multi-step type, equipped with an endless chain connected by a wire net is classified into Six, Eight and Ten steps. This type of dryer is vertically divided into three sections: upper, middle and lower each having a different temperature. This type is suitable for a large-scale reeling factory, which has a drying capacity of 5 000 to 18 000 kg of fresh cocoons per 24 hours.

— *One band type :* This type has one band of cocoon feeding wire net which carries at least one foot of cocoon piling height for drying, while the Multi-step does only 5 cm of cocoon piling height on the cocoon feeding wire net. There is no difference in the drying capacity between both Multi-step and One-band types. The latter needs less space for installation, but it consumes more electrical power for the drying operation.

c) Low temperature dryer. This dryer has a dual purpose of cocoon drying and storage. A continuous supply of fresh cocoons is needed for completion of one operation. More attention should be given to the operation of the dryer, especially temperature control.

Sorting Machine

A small-scale reeling factory is equipped with a simple type of sorting table. A large-scale factory is equipped with a conveyor or system attached to the sorting machine.

Cooking Machine

The proper capacity of the cooking machine should be selected, since most cooking machines are basically the same in structure. In addition, the devices for aiding the pre-treatment part and cooking control parts should be considered for use, depending upon the quality of cocoons, reelability and cocoon shell percentage.

Reeling Machine

The reeling machine to be selected usually depends upon the quality of the available cocoons.

a) *Simple reeling machine.* This type of reeling machine has five to six ends per basin without the Stop motion device. For this machine, the operator is seated and processes poor quality cocoons. The reeling capacity is in the range of 500 to 600 g per basin per 8 hours.

b) *Multi-ends reeling machine.* The Multi-ends reeling machine is equipped with 20 small reels per basin. The groping ends, picking ends and cocoon feeding of reeling thread is done manually with the reelers in a standing position. The reeling capacity of this machine is in the range of 800 to 1 000 g per basin per 8 hours. This equipment may be used with all qualities of cocoons.

c) *Automatic reeling machine:* An automatic reeling machine may be used effectively if good quality cocoons with 60 percent reelability, 1 000 m filament length and 20 percent cocoon shell are used. In automatic reeling, the production of raw silk can be increased with less labour, compared to Simple type or Multi-ends type reeling.

Re-reeling Machine

The re-reeling machine is composed of large wheels known as "window", which reel five or six skeins per wheel. One window of five skeins wheel has the re-reeling capacity of ten skeins per 8 hours, if one skein weighs around 130 g. It is possible today to find the High-speed re-reeling machines on the market.

By-product Treatment Machine

On an industrial scale, the reeling factories in excess of 1 000 ends require installation of a by-product treatment machine to separate pupae from parchment layer "reeled out cocoons". This function is done manually in small-scale factories.

Reel Permeation Device

The reel permeation device "Vacuuming treatment tank" is necessary for the pre-treatment of small reels for the re-reeling process. This device is especially useful for double skein reels, which have over 70 g of raw silk wound on to a reel.

Skein Twisting and Booking Machine

There are two kinds of skein twisting and booking machines. One is the short skein-booking machine by which the skeins are twisted and folded once. Another is the long skein-booking machine by which the skeins are twisted on the turntable and booked without folding.

Raw Silk Testing Equipment

The raw silk testing equipment for quality control in the reeling factory is required as follows:

- — Seriplane machine
- — Seriplane illumination equipment
- — Conditioning oven
- — Denier scale and Sizing reel
- — Rewinding machinef

The Serigraph and Cohesion testing machines are not necessary for reeling factories, but are for quality assurance and research institutions

Table 1. Items of machinery required for a silk reeling factory (Capacity 70 M/T fresh cocoons yearly)

Types of machines	*Specifications*	*Quantity*	*Capacity*
Cocoon drying machine	Hot air blowing and circulating system, three horizontal sections upper, middle and	1 set	600 kg of fresh cocoons per 8 hrs.

	down rooms with automatic electrical thermostats L 12 700 x W 5 300 x H 2,000		
Cocoon cooking machine	Endless chain system. 1 No. of baskets : 70 L 8 550 x W 1 190 x H 1,795		200 kg of dried cocoons per 8 hrs.
Cocoon reeling machine	Semi-automatic type with denier detecting system 14 ends per basin L 16 135 x W 1 360 x H 1 795	32 basins	117 kg of dried cocoons per 8 hrs.
Silk re-reeling machine	5 heads per wheel L 18 470 x W 1 360 x H 1 360	40 wheels	45 kg of raw silk per 8 hrs.
Vacuum permeation device	Power : 1.5 hp ? 750 x H 1 200	1 set	100 kg of raw silk per 8 hrs.
Cocoon assorting machine	Conveyor type	1 set	300 kg of dried cocoons
By-product treating machine	Automatic type L 7 880 x W 545 x H 1 000	1 set	30 kg of dried pellete per 8 hrs.
Silk booking machine	Motor driven	1 set	
Silk skein twisting machine	Hand driven	1 set	
Seriplane winding machine	20 black boards, carrier and Standard Photos included	1 set	
Illumination equipment for inspection	For Cleanness and Neatness test	1 set	
Denier scale	400 revolution and 200 revolution	2 sets	
Sizing reel	Circumference of reel 1.125 m	2 sets	
Rewinding test	10 bobbins	1 set	

Contd....

Contd....

machine		
Boiler	1	2 tons
Oil tank	1	
Water tank	1	
Chimney	1	
Electric equipment	1	
Mechanical tools	1	

Table 2. Building space required for each item of machinery to be installed

Rooms	*Space (m^2)*	*Remarks*
Cocoon drying	165	Hot air circulating type : 1 set
Cocoon storage room type	165	Honey comb tube type and general
Cocoon sorting set	30	Conveyor type sorting machine : 1
Cocoon cooking	60	Auto cocoon cooking machine : 1 set
Reeling basins	230	Semi-automatic reeling machine : 32
Re-reeling	165	Re-reeling machine "32 wheels
Inspection and finishing	40	Seriplane winding machine, illuminating equipment and rewinding test machine
By-product treating	60	Auto by-product treating machine : 1 set
By-product storage	30	
Boiler	50	Boiler 1 set (2 tons)
Mechanic	30	Tools and spare parts
Raw silk storage	30	
Rest	50	
Dormitory	100	
Officials	50	
Total	1,255 m^2	

13

Simulation of Reeling Unit

Silk is a crazy business owing to its long standing reputation of being risky and mentions that profits fluctuated a lot. The approximate cause of these fluctuations has been the fluctuations in the price of silk in the period between the harvest of cocoons and the sale of silk wares to the consumer.

There is a wide spread complaint from the reelers that reeling activity is a losing enterprise. The lack of transparency on the part of the reelers makes the reeling activity more or less a black box. A few studies carried out on the economics of silk reeling industry have also indicated that reeling is largely a loss making activity. Thus there is a need to have a thread bare understanding of the functioning of a typical reeling unit, the fluctuations in the cocoon and raw silk prices and in turn its influence on net revenue and utilisation. Seasonality in cocoon supply, cocoon quality as also its price, are quite well known.

A reeling unit, which is subject to these vagaries would certainly have changing fortunes. It is, therefore, of utmost importance to study the operations under seasonal differences over a whole cycle, to gather their cumulative effect. Since profit in reeling is expected to be a function of cocoon and raw silk prices, which experience wide fluctuations with certain associated probabilities, a dynamic study of the functioning would alone reflect the performance of a reeling unit. Lamm highlights the usefulness of simulation as a viable and powerful method for evaluating market performance especially when industry accounting information is not readily available. This exactly echoes the constraints of any detailed study in respect of Indian silk industry. Charsley refers to the peculiar problem of our silk industry. He states that the lack of any documentation or accounting in most of the industry and hostility to financial enquiry among the few who may undertake some accounting for their own benefit, makes it extremely difficult to obtain the data necessary for financial evaluation.

To assess the annual cumulative net revenue of a typical reeling unit, a study was conducted in Karnataka recently. In order to simulate the operations of a reeling unit over a period of one year, to capture the vagaries of seasonality, it was necessary to build a valid model capable of truly representing a typical reeling unit. Since a wholly analytical modeling was considered unsuitable due to the complexities involved, it was decided to capture the heuristic thinking of a few reelers and to develop a heuristic decision model to reflect the practices ingrained in the reeling units and of strategies they adopt to counter the vagaries of their environments and operations.

Accordingly, as the first step, a heuristic model of a typical reeling unit was developed, considering the various input parameters and the environmental conditions. Reeling being an input controlled industry, the heuristics of cocoon buying decision was incorporated as the core of the model. The heuristics were derived from discussions with several prominent reelers and observation of their operations .

In the second step, the model was used to simulate its working over a period of one year. The sheer complexity of the system of interest and of the models necessary to represent them in a valid way, precludes the use of analytical solutions and make simulation the right choice for numerically exercising the model for the inputs in question to see how it affects the output measures of performance.

Discrete, dynamic and stochastic simulation models are called discrete-event simulation model. Discrete-event simulation concerns the modeling of a system as it evolves over time by a representation in which the state variables change instantaneously at separate points in time i.e. the system can change at only a countable number of points in time. These points in time are the ones at which an event occurs.

Simulation Model Building

Leung et al observe that building of simulation models has long been considered an art rather than a science and there is no systematic approach. He further adds that a modeler can generally understand a problem through interviews, questionnaires, observation and existing documents. According to them, simulation model building requires to define the scope of the system, build a list showing all the events that can cause a system to change its state and finally to show how to model the entities, activities and time sequences of events of a system.

The primary objective during this phase of model

building is to develop a model with high face validity, i.e. a model that on the surface seems reasonable to people who are knowledgeable about the system under study. In order to develop such a model, the simulation modeler should make use of all existing information that include: conversations with system "experts"; observations of the system; existing theory; relevant results from similar simulation models; experience and intuition. They zero down on two ideas as the most important part of model building. First is interaction between the modeler and the manager/user throughout the course of the simulation study. Another very important idea for validity/ credibiHty enhancement is for the modelers to perform a structured walk-through of the conceptual model before an audience of all key people.

The best compass for modeling is a well-defined set of objectives. This is because the objectives orient the project by providing criteria against which modeling decisions can be judged. Next, he says, one should determine what output measures are needed to answer these objectives. Then, by focusing on these measures and only these measures, the model's scope can be more narrowly defined and its detail reduced, he concludes. The sources of information can be overwhelming or non-existent. Data sources, they say, include interview with experts, time studies, manual data records, computer logs, personal observations, documented procedures etc.

Heuristic Model

When the mathematical model of a problem is of such a nature that an analytical (closed form) solution procedure is unknown, a heuristic method is used. It is often used for a well defined problem that can be used for learning purposes to develop an intuitive feeling as to important variables. This closely parallels primary reasons for simulating a system. Realistic formulations of complicated problems lead to mathematical problems very difficult to solve and the use of heuristic methods assumes importance.

In heuristics, one endeavours to understand the problem solving process especially the mental operations of a human problem solver. Heuristics are rules of thumb that allow us to factor approximately the complex perceived world into highly simple components and to find approximately and reasonably valuable solutions that allow us to act on that world predictably. According to Merbach, heuristics is usually understood "as an iterative algorithm which converges toward the solution of a problem".

Heuristic programming utilises rules of thumb or intuitive rules and guidelines and is generally under computer control to explore the most likely paths to take and to make educated guesses in arriving at a problems solution. It draws from the knowledge and experience of the system in which the problem resides. Heuristics have been extensively used in stock market analysis and accounting. There are many management oriented heuristic applications to problems such as assembly line balancing, facility layout, portfolio selection, job shop scheduling warehouse location, inventory control, resource allocation etc. Of the many types of heuristics developed for problem solving, particularly in business and industrial situations, two are of special interest to the current application;

(1) Construction method in which heuristic rules are used in a sequential, deterministic manner and there is a local emphasis and

(2) Decomposition methods in which the problem is broken into smaller manageable parts and the solutions of the components are integrated.

One of the major problems of heuristic modeling is that evaluation of the heuristics is difficult, particularly when an optimal solution is not available. In many cases, comparison with the results of the actual system and face validation by experts may have to be used, particularly when complex systems are being modeled through simulation, primarily to be understood.

Data for Model Building

In an analytical study of the silk reeling operations in Karnataka, B.V Vasumathi adopted the above approach for model building. The details of her simulation experiments are described below.

— Reeling units in three traditional clusters of Karnataka viz., Bangalore rural Mysore and Kolar districts were extensively visited. Detailed discussions were held with reeling unit owners trying to understand their problems in the operations of the unit as also constraints faced by them. Details of operational cost, labour cost, raw material availability, cycle time, working capital availability etc were gathered. Around 120 reelers, 40 from each district were interviewed for a surface understanding.

— Later, owners of a handful of multiend reeling units in the state were interviewed for collecting operational details in depth, especially the aspect of buying cocoons from the markets.

— CSTRI Project Report for different multiend reeling units with projected cost, production, input, output and sales details also formed one of the important sources of information for model building.

— Discussions with reeling experts in CSTRI as well as or specialists elsewhere contributed positively to the process of model building.

— The experience and intuitive understanding of the industry by the author came in handy, in building a credible model.

Basic Blocks of the Model

The functioning of a reeling unit is largely dictated by its cocoon purchase decision, which decides the key issues such as cocoon consumption, raw silk production, capacity utilisation, through put etc. Since the cost of cocoons constitute 80-90% of the cost of raw silk production, the cocoon purchase decision significantly decides the revenue of the reeling unit. The cocoon purchase also has a critical bearing on the quality and quantity of raw silk reeled. Thus, cocoon purchase decision comprising the aspects of how much to buy, of what quality, at what price and when to buy would determine the fortunes of a reeling unit.

There are four basic blocks of considerations in the reefer's mind about the operations of the reeling unit.

(a) To find out whether he can generate funds by selling raw silk to meet the requirements of working capital and still maintain a minimum raw silk inventory for better realisation. A notional upper limit for the quantity of raw silk in inventory is fixed.

(b) To keep down the inventory of cocoons but still be able to effect opportunity purchases and run the reeling unit with minimum interruptions.

(c) To find out whether the cocoon market conditions are favourable for purchase of cocoons—this is achieved by comparing the notional perceived price and actual price of cocoons.

(d) To decide whether to purchase and how much to purchase depending upon

— The extent the cocoon market is unfavourable.

— The level of cocoon inventory available.

— The trend of cocoon availability and in turn its price.

— Whether the basic costs of production can be recovered.

— The availability of funds.

Operational Strategies

As per the method adopted by Bailey in selecting the marketing strategies used by cotton producers, the operational strategies of the reeling units were identified based on the interviews with manufacturers, traders and experts in the silk industry.

— *Strategy 1*- Reeling unit operating with a prior commitment of supply to the buyer, which he has to honour up to a limit as restricted by the working capital and price situation.

— *Strategy 2*- Reeling unit operating with compulsions to retain the skilled labour.

— *Strategy 3*- Reeling unit operating without any constraints and is free to temporarily withdraw from operations whenever the conditions are not so favourable.

These strategies are incorporated in the development of the model.

Events and Variables of the Model

A number of variables and parameters related to a reeling unit were identified during the analysis phase for model building. They are listed below:

Input Variables

1. Cocoon Quality Index
2. Cocoon Price
3. Raw Silk Price

Events

1. Sale of Raw Silk
2. Production Scheduling
3. Cocoon Evaluation
4. Purchase of cocoons

State variables

1. Status of Plant
2. Number of Operational Basins
3. Cost of Cocoons Consumed
4. Raw Silk Sale Price
5. Raw Silk Production Per Day
6. Cocoon Inventory
7. Cocoons Purchased
8. Raw Silk Inventory
9. Raw Silk Sold

Response Variables

The model outputs, typically representing the performance of the system are:

1. Cumulative Capacity Utilisation
2. Cumulative Net Revenue

These two measures of performance are discrete-time statistics, since it is defined relative to the collection of random variables (cumulative revenue and capacity utilisation) that have a discrete 'time' index, i = 1, 2, 3, —, 365.

Heuristics

There are two main heuristics developed for the purpose of incorporating the operational thinking of the reelers. They are (1) Cocoon price foresight heuristics and (2) Cocoon evaluation heuristics, which are described below:

Cocoon Price Foresight Heuristic

An intelligent reeler has his antenna open to absorb the signals of the trade and is adequately cautious. It was, thus, felt very important to map these various perspectives of cocoon purchase so as to evolve a comprehensive approach and make the model as realistic as possible. In this pursuit, it was proposed to encapsulate the knowledge base of the reeler to identify the alternatives for the functioning of a reeling unit.

The reeler tends to keenly observe the fellow reelers' assessment of the cocoon lot, which bolsters his own assessment of the cocoon lot that he intends to buy. The reeler also makes his own forecast of the quantity flow of cocoons in the market, based on the information he has of the eggs purchased by the farmers in the area, the rainfall and the mulberry transacted. The model attempts to include this foresight as a variable defined as TREND which can be positive, negative or neutral. For calculation of TREND, in order to use in the simulation experiment to be described later, the weekly average cocoon price was compared with the monthly average price. If there existed a significant difference between them, then weekly average price was compared with the average price of the subsequent week. If the average price of the subsequent week was significantly higher, TREND was taken as positive, if it was significantly lower, TREND was taken as negative, else it was considered neutral. Simple Y test was used to compare the means. The TREND values were read from an input file consisting of a table depicting day wise trends.

Cocoon Evaluation Heuristic

The model tries to incorporate the volatile situation of the cocoon market and the raw silk sale price as also the constraints of working capital and the decision the reeler tends to take under these conditions. Attempts have been made to map the heuristics of buying decision that goes on in a reeler's mind and the decision he ultimately takes with whatever little information he has at his disposal.

The reeler normally visits the cocoon market every day, though he may choose to purchase on only 4 or 5 days in a week. He initially, subjectively assesses the quality of the cocoon by visual and tactual methods and estimates what could be the yield obtainable from these cocoons. Based on his estimate of the renditta and the prevailing raw silk price or even considering the forecast of raw silk price in the coming week, he further estimates the kakame cost. His experience with the conversion and labour cost involved and the profit he expects are also considered for the purpose.

From the estimated kakame cost and renditta, he arrives at the price payable to the cocoons, which will be referred to as the perceived price of cocoons in the model. He compares this with the actual price of that particular lot of cocoons, which enables him to understand whether the market condition is favourable to make the purchase.

The conditions are termed favourable when actual price of cocoons is less than the perceived price. Here the reeler makes a decision to buy cocoons irrespective of whether the TREND is positive or negative, and TREND affects only the quantity of purchase. The capacity to store the cocoons, cocoon inventory and the availability of funds also contribute to the decision of how much to buy. When the perceived price of cocoons is less than the actual price, the conditions are termed unfavourable and the decision of the reeler to buy or not to buy and how much to buy varies depending on the strategy of operation.

These two heuristics were suitably incorporated in the form of simple rules into the decision model of the reeling operations and are depicted in a logical manner in the flowcharts of the model.

Model Assumptions

(1) Profit is not ploughed back into the business.

(2) The raw silk is transacted in terms of cash payment and there is no delay in realising the payments.

(3) The market demand for raw silk readily exists.

(4) The production efficiency is the same in a given season.

(5) The quality of raw silk produced is in the average quality level of a known range, fetching a price in the average price level.

(6) Scope for opportunity purchase exists, while there is no scope for opportunity sales.

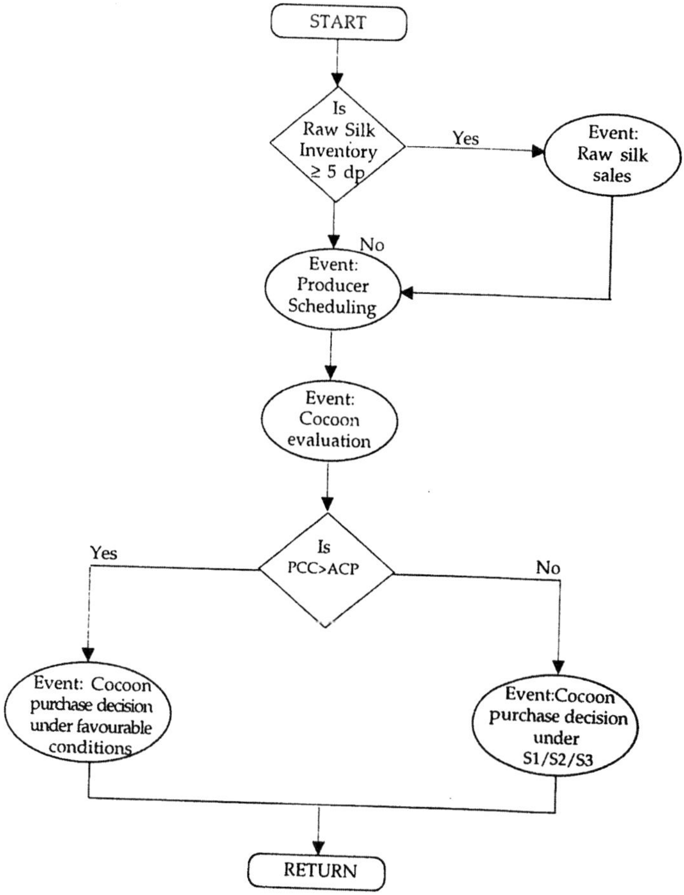

dp = standard raw silk production per day
PCC = Percieved cost of cocoons
ACP = Actual price of cocoons
S1 (strategy 1) = Reeling unit with commitment to supply raw silk
S2 (strategy 2) = Reeling unit with commitment to retain the skilled labour
S3 (strategy 3) = Reeling unit without any constraints

Figure 1. Heuristic model of a reeling unit's operations

(7) Borrowed money is made available only for the purpose of cocoon purchase.

(8) Working capital is available to the reeling unit at bank rate of interest of 16.57c per annum.

(9) Variable costs only have been considered in the model and fixed costs have not been considered.

(10) There is no limit for the availability of skilled labour.

(11) On weekly holiday, production does not take place and raw silk cannot be sold, but cocoons can be purchased.

(12) Except on two national holidays, the cocoon transactions in the markets take place daily.

(13) The reeler has adequate experience and skill to make a perfect estimation of cocoon quality, during purchase.

Heuristic Model of Reeling Uni Operations

The computer model of the daily operations of reeling unit incorporates the details of the heuristics discussed above. The overall model is succinctly portrayed in Figure 1 and the sub models are detailed from Figures 2 to 8. These are referred at appropriate points in the discussions.

The three events viz., cocoon purchase, production scheduling and raw silk sales are almost parallel activities on a given day. However, the day's production of raw silk is not available for sale on the same day, the returns from raw silk sale on a given day is not available for effecting cocoon purchases on the same day, and the day's ɔcɔon purchase is made available for production with a lag of two days. These aspects have been the guiding factors for sequencing the events.

A typical reeler uses the following heuristic thinking in the day to day operations of his reeling unit.

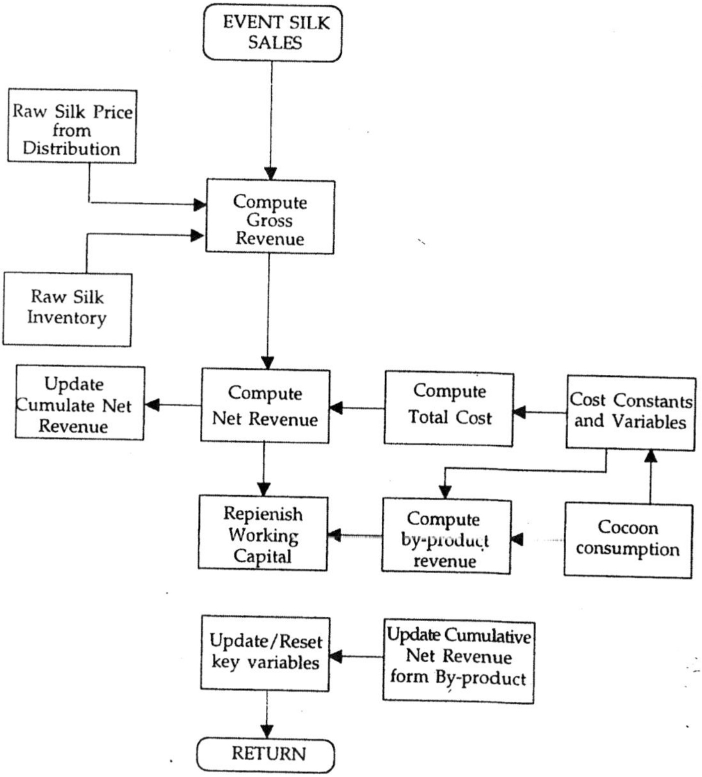

Figure 2. Flowchart for raw silk sales

Sales decision

If the inventory of raw silk exceeds a certain predefined limit, he sells the raw silk and then proceeds to schedule the plant for production. If the raw silk inventory is less than the limit, then he will straight away proceed to production scheduling. When the raw silk sales occur, revenue details are computed and working capital is replenished.

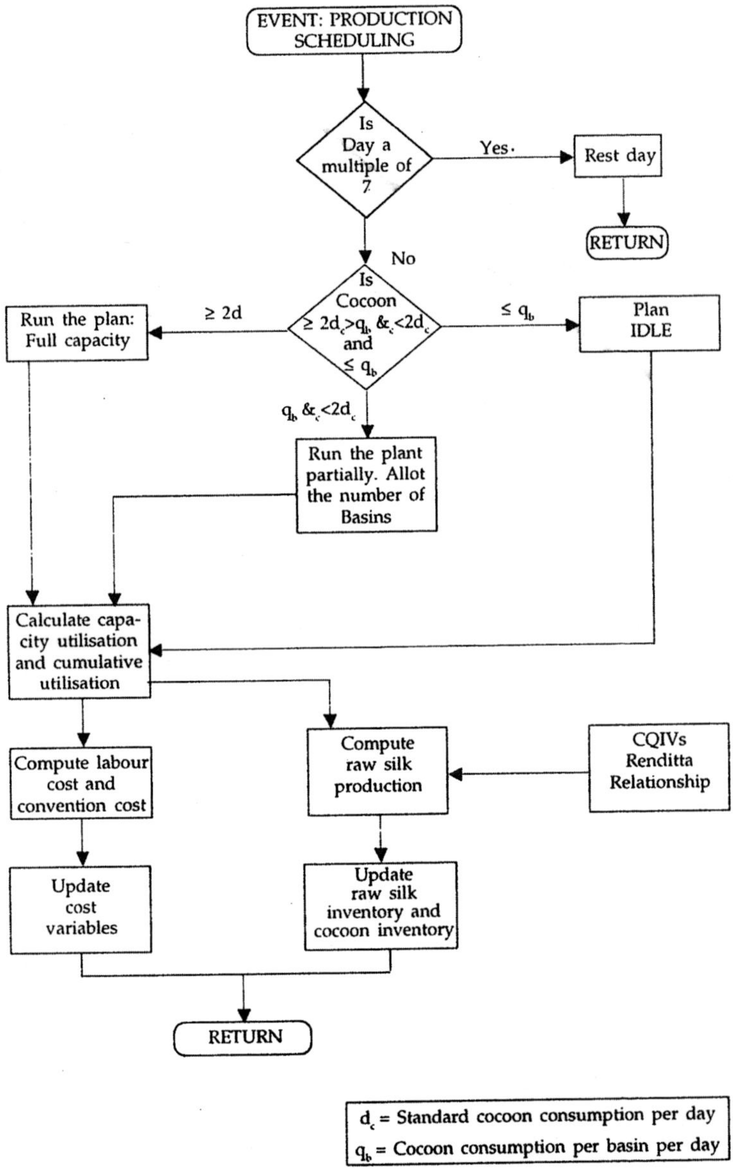

Figure 3. Flowchart for production scheduling

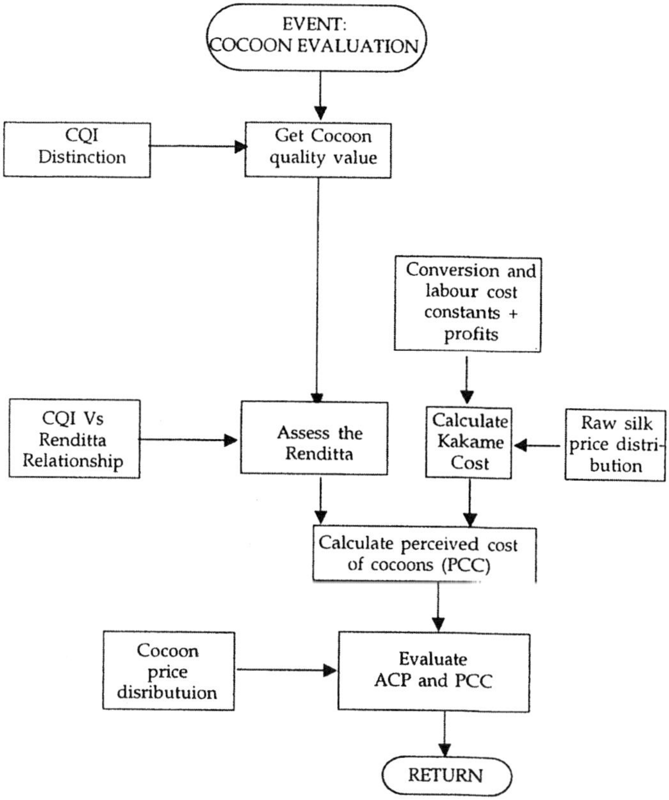

Figure 4. Flowchart for cocoon evaluation

Production scheduling decision

Depending upon the cocoon inventory available, the reeler decides to:

— Run the plant fully, if the cocoon inventory available is more than a predefined maximum value.

— Run the plant partially, if the cocoon inventory available is between a maximum and a minimum value.

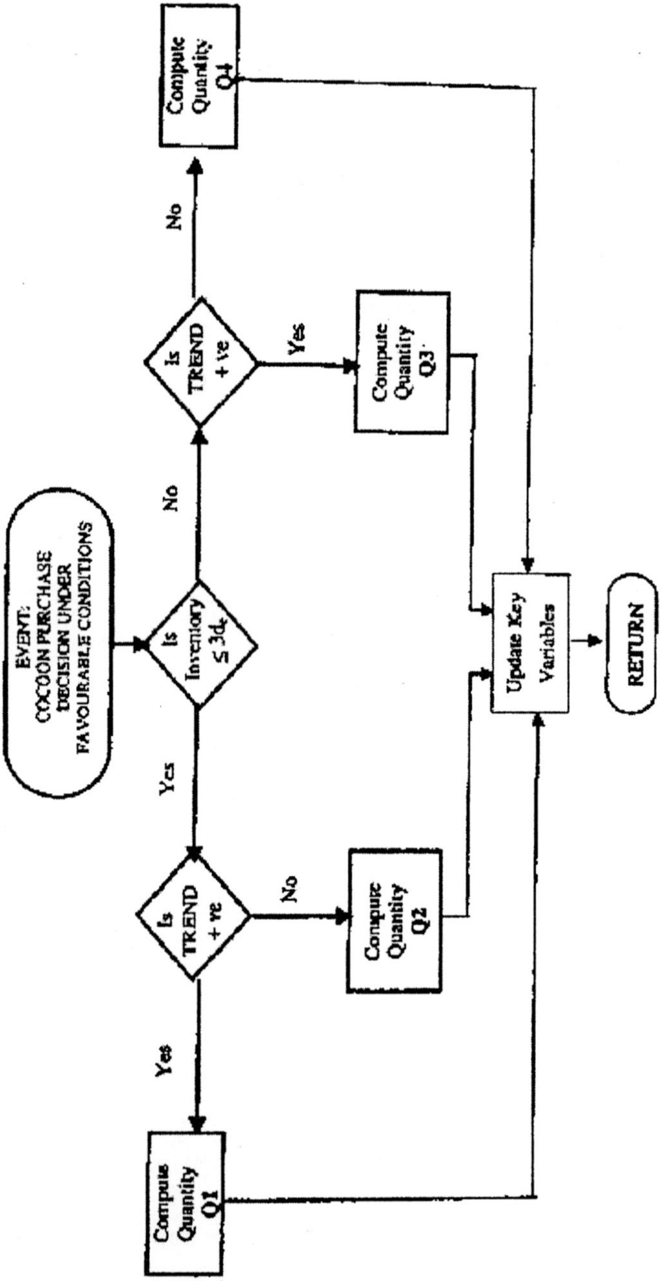

Figure 5. Flowchart for cocoon purchase under favourable conditions

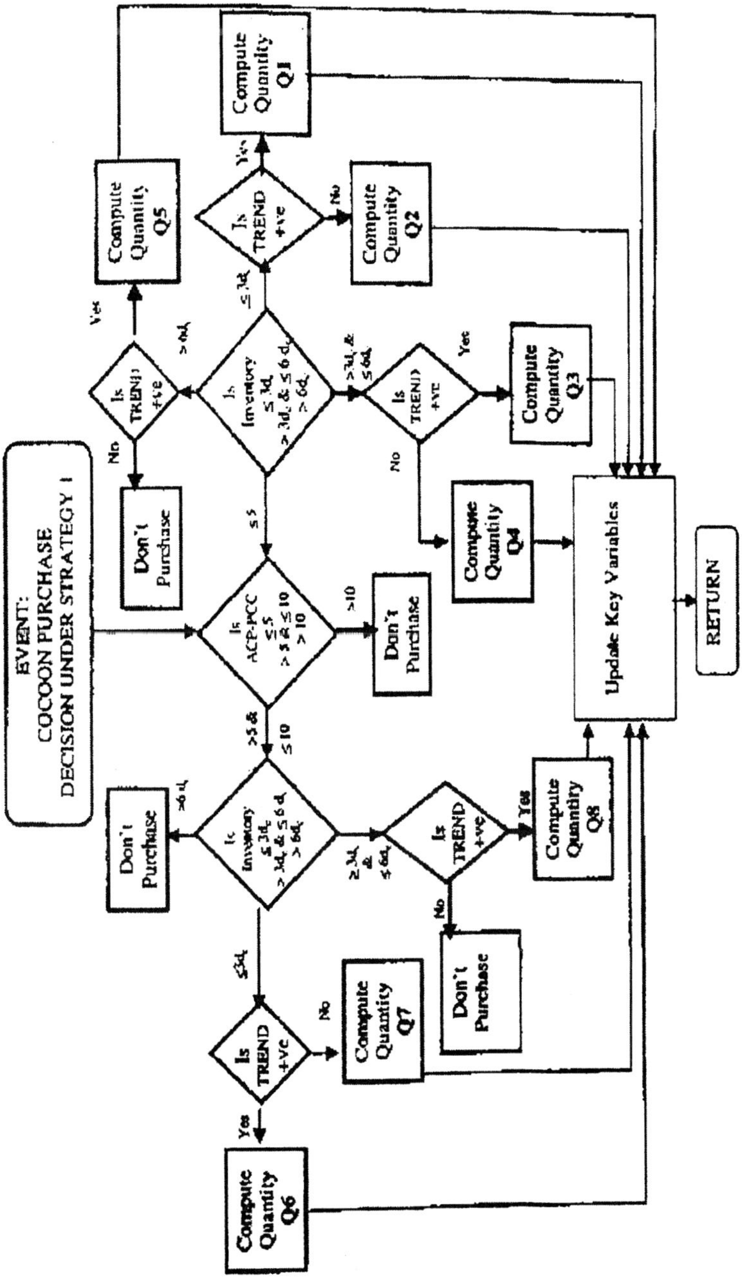

Figure 6. Flowchart for cocoon purchase decision under strategy 1

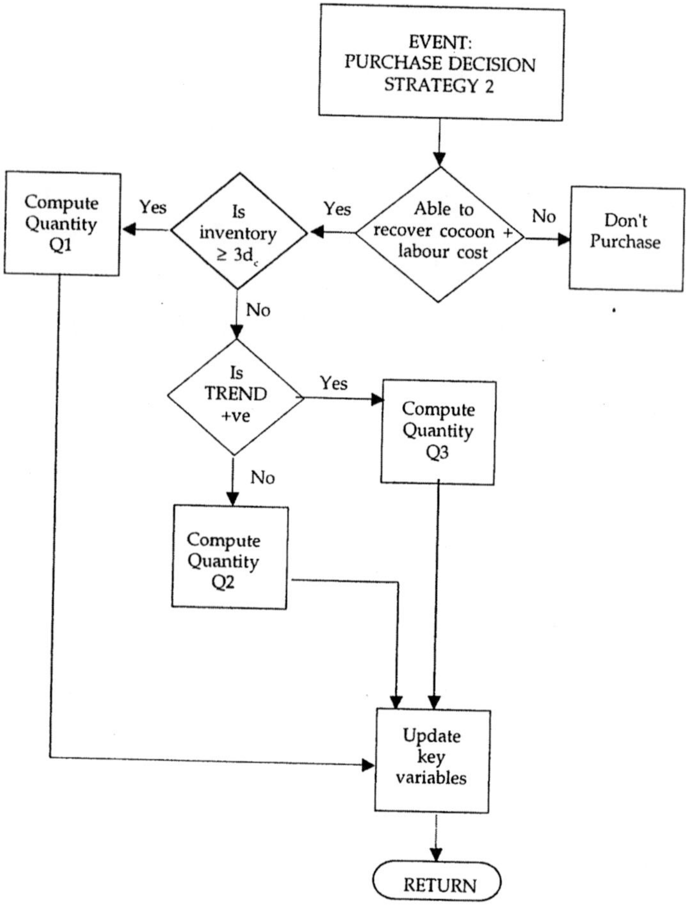

Figure 7. Flowchart for cocoon purchase decision under strategy 2

— Keep the plant idle, if inventory available is less than the minimum value.

The cocoon and raw silk inventory levels are updated and watched.

Cocoon evaluation for purchase decision

Purchasing decisions are made depending upon two issues:

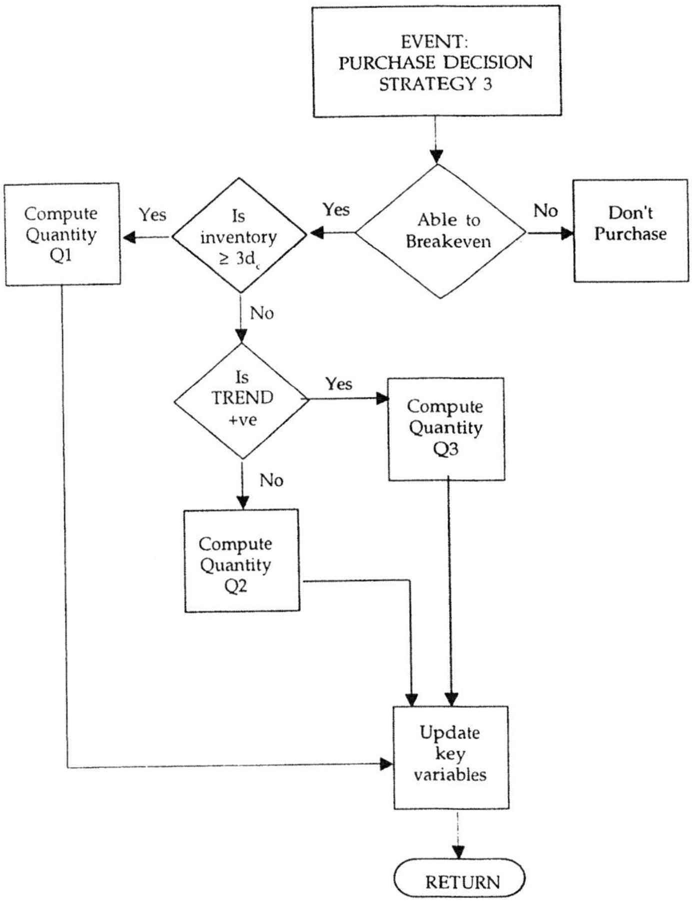

Figure 8. Flowchart for cocoon purchase decision under strategy 3

— Whether the conditions are favourable for purchase or not.

— When the conditions are unfavourable, what is the strategy adopted by the reeler.

The heuristics to evaluate cocoons and then decide whether the conditions are favourable or not is a very ingenuous one and is referred to as cocoon evaluation heuristics. When cocoon purchase decisions are made, care has been taken to ensure that the inventory did not exceed the limit set on the inventory.

(i) *Favourable conditions.* The decision depends hierarchically upon two issues:

— Whether the inventory of cocoons available is greater or less than a stated number of days consumption.

— Whether the TREND is positive or not.

Here, scope fur opportunity purchase exists. There are four purchase quantity decision options and the reeler tends to make purchase decision unless otherwise restricted by the working capital.

The purchase quantities as indicated in the compute quantity boxes in the flow chart are given by the following options:

Q1 = Min <lOdt, WC/ACP)

Q2 = Min (6.5 do WC/ACP)

Q3 « Min { NPQ, WC/ACP)

Q4 = Min (MPQ, WC/ACP)

Whem d_c = Standard cocoon consumption per day

NPQ se Normal Purchase Quantity

MPQ = Medium Purchase Quantity

LPQ as Low Purchase Quantity

WC « Working capital available

ACP = Actual cocoon price

(ii) *Unfavourable conditions.*

Strategy 1. The decision depends hierarchically upon three issues:

— Whether the difference between perceived price and actual price of cocoons is greater than or less than pre stated levels.

— Whether the inventory of cocoons available is greater or less than a stated number of days consumption.

— Whether the TREND is positive or not.

There are eight purchase quantity decision options and five options of not purchasing, depending on the situation in this strategy. The quantities indicated in the compute quantity boxes in the flow chart are as follows:

— Q1 = Min (NPQ, WC/ACP)

— Q2 = Min (MPQ, WC/ACP)

— Q3 = Min (MPQ, WC/ACP)

— Q4 = Min (LPQ, WC/ACP)

— Q5 = Min (LPQ, WC/ACP)

— Q6 = Min (MPQ, WC/ACP)

— Q7 = Min (LPQ, WC/ACP)

— Q8 = Min (LPQ, WC/ACP)

The inventory level is categorised much finer here since the reeler tries to avoid purchasing under unfavourable conditions to the maximum possible extent, while at the same time, embarking on continuous production without breaks. A minimum of three days inventory is what ensures this.

Strategy 2. The decision here depends on three issues in a hierarchical order:

— Whether the cost of cocoons and labour costs can be recovered, if the cocoons are purchased at the bid price.

— Whether the inventory of cocoons available is greater or less than a stated number of days consumption.

— Whether the TREND is positive or not.

There are three purchase quantity decision options and a no purchase decision under this strategy. The quantities indicated in the compute quantity boxes in the flow chart are as follows:

— Q1 = Min (MPQ, WC/ACP)

— Q2 = Min (LPQ, WC/ACP)

— Q3 = Min (MPQ, WC/ACP)

The cocoon inventory is categorised into only two levels since there is a stringent condition to be met for decision to purchase and finer classification of cocoon inventory becomes redundant. Here the reeler is ready to forego the conversion cost since there is a cost associated with non-utilisation and he may be better off losing this cost than paying idle wages.

Strategy 3. In this decision, there are again three issues considered in a hierarchical order.

— Whether the cost of cocoons, labour cost and conversion cost can be recovered.

— Whether the inventory of cocoons available is greater or less than a stated number of days consumption.

— Whether the TREND is positive or not.

There are three purchase quantity decision options and a no purchase decision under this strategy. The quantities indicated in the compute quantity boxes in the flow chart are as follows:

— Q1 = Min (NPQ, WC/ACP)

— Q2 = Min (LPQ, WC/ACP)

— Q1 = Min (NPQ, WC/ACP)

With an inventory level of less than or equal to 3 days cocoon consumption, here the reeler does not look at the TREND, since he tries to purchase minimal quantity sufficient to keep the unit running and thus the TREND over the subsequent week becomes out of context here. It is very much relevant to note that, most reelers always choose the cocoons in a particular quality range and do not switch between the extreme qualities. Once the quantity of cocoons purchased is generated in simulation, the production and sales of raw .silk and ultimately the net revenue and capacity utilisation of the reeling unit will be simulated, as described in the relevant sub models.

Discussion of Model Parameters

CQI and Renditta

CQI and the corresponding renditta were collected from a multiend reeling unit for a period of two years. The values were plotted. Two slopes were observed, one in the lower range and one in the upper range of CQL Regression analysis was carried out for cocoon quality index and renditta separately for the two ranges, to fit two different equations for estimating renditta from CQI values. This would take care of the renditta difference experienced in the reeling units in the favourable and the unfavourable seasons.

Seasonally

Experience as well as the views expressed by the reeling unit owners indicated that the going is good in the months starting from January upto June. This is because, good quality cocoons are available during this period, ensuring good yield and productivity. As the rainy season sets in and till almost the end of winter, the quality of cocoons deteriorates. The conversion cost for fuel and water too goes up, since the fuel requirement shoots up owing to the dampness. The basin water needs frequent changing since water gets turbid much faster owing to the poor quality of cocoons. Though in reality January, July and August are more transitory in nature depending on the onset of monsoon and the phasing out of winter, for the purpose of modeling, January has been included in favourable season consisting of 6 months from January to June, The later months from July to December have been categorised as unfavourable season. The yield i.e., renditta, labour and basin productivity as well as conversion cost have been accordingly considered.

Cocoon inventory

An initial cocoon inventory of 3 days consumption was taken in the model so as to ensure continuity in production, as there exists a lag of 2 days in using the cocoons purchased. A limit on the cocoon inventory was fixed at 12 days consumption. The day to day cocoon purchase gets into the inventory and cocoons are drawn from the inventory for production. Thus, weighted average cocoon price and weighted average cocoon quality were computed with each lot of cocoon purchased. This was considered for cost calculation and also for calculating renditta. The aspect of two days lag between purchase and consumption was duly taken note of, in updating the weighted average cocoon price and quality. Working capital was updated with each lot of cocoon purchased, The cost of inventory was not explicitly considered in the model, but was taken care of, by the interest component of the working capital and borrowed money.

Working capital and borrowed money

For reeling units, the start-up is at times relatively easy because it can he done on a very small scale, but working capital is always, even on the smallest scale a problem.

The reelers have a practice of using money borrowed from private lenders at a high rate of interest. This was considered as borrowed money in the model. This borrowed money will be made available strictly for the purpose of cocoon purchase. It is normally taken when working capital is insufficient and there are compelling reasons for the purchase of cocoons. Generally up to 25% of one's working capital used is borrowed money, which has been incorporated in the model.

Cost parameters

The cost components considered in the model were the cost of cocoons, labour cost, conversion cost (other costs consisting of fuel electricity and water expenses), market fee of 1 c/c and interest on working capital and borrowed money. The revenue from the main product namely raw silk alone was considered for calculating revenue. The income from the by-products viz., defective cocoons and silk waste was taken separately.

The cycle time was taken as 5 days production i.e., a quantity of 50 kg of raw silk for a 10 basins unit. There is no standard quantity for sales in the industry and it generally varies from 30 to 40 kg. It was felt necessary to bring in a certain amount of orderliness and therefore a notional batch size of 50 kg for transaction, was considered appropriate in the model for the main analysis. However, for the purpose of comparison, runs were also made with a batch size of 30 kg of raw silk.

When the net revenue at the end of cycle is positive, the total cost component was put back into the working capital fund earmarked for cocoon purchase, in the ratio of 6:1. One seventh was kept as cash on hand for meeting the day to day expenses other than cocoon purchase. When the net revenue at the end of a cycle was negative, the net revenue component was put back in the ratio of 6:1, for the purpose of cocoon purchase. However this ratio turned out to be 8:1 in the unfavourable season. The ratio was decided based on the proportion of cocoon cost to all the other costs on a day to day basis.

By-products

5 per cent of the cocoons consumed was taken as cocoon waste, which fetched an average price of Rs 75/- per kg. The silk waste recovered was taken as 30% on the weight of cocoons consumed. The silk waste fetched an average price of Rs 120/- per kg.

Cost of non-utilisation

Reeling units operating with a compulsion to provide work to the skilled labour on a continuous basis are required to pay 50% of the wages, whenever a minimum of 50% of the labour force are not engaged on any given day. This component was added to the total cost of production. Cost of non-utilisation in the units operating under the other two strategies, were considered as sunk cost, since there was no other explicit cost component in these cases.

Cocoon purchase quantity

In the normal course, 3-4 days cocoon consumption is purchased at a time., which has been defined as Normal Purchase Quantity in the model. In addition, Medium Purchase Quantity and Low Purchase Quantity have been defined as 2-3 days and 1-2 days cocoon consumption, respectively, in the model.

14

Production and Marketing of Quality Cocoons

Cocoons, the output of the rearing sector forms the input for the reeling industry. The silk cocoon consists of the outer cover which is the shell comprising continuous filament spun by the silk worm and the pupa inside, besides a lot of air space between the shell and the pupa. The moisture in the shell and pupal body is generally to the extent of 60-64% of the weight of the green cocoon. The silk content in the cocoon is expressed in terms of shell ratio percentage. Higher the silk content in the cocoons, higher will be the yield of raw silk from a given quantity of cocoons.

The quality of cocoons has a significant impact on the yield of raw silk, which is expressed in terms of renditta (number of kg of cocoons needed to reel one kg of raw silk) and quality of raw silk reeled and reeling filature efficiency. Cocoon quality contributes to the tune of 80% towards raw silk quality and productivity. Further, cocoons as a raw material have a significant bearing on the reeling industry since it contributes 85-90% of the cost of production of raw silk.

Cocoon quality plays a decisive role in the reeling performance and quality of raw silk. The value of cocoons as a raw material for silk production depends to a very great extent on its raw silk recovery percentage. The cocoon filament length, reelability of cocoons and incidence of defective cocoons in the lots have a direct bearing on the silk percentage of cocoons. The occurrence of more defective cocoons in the cocoon lots will affect the economics of reeling and quality of raw silk.

Cocoon Quality Parameters

Cocoon quality is governed by the following parameters, each of them being associated with a certain level of importance:

(1) *Shell Ratio Percentage (SR%):* This is the ratio of the shell weight to the cocoon weight of a given number (usually 50) of cocoons expressed as a percentage. Higher the SR%, better is the quality of cocoons.

(2) *Defective Cocoon Percentage (DC%):* This is the total number of defective cocoons (like double cocoons, flimsy cocoons, melted cocoons, infected cocoons, urinated cocoons and malformed cocoons) in one kg of cocoons expressed as a percentage of the total number of cocoons in it. Lower the DC%, better is the quality of cocoons.

(3) *Average Filament Length (AFL):* It is the total length of the filament in metres that can be fully unwound from a single cocoon. It is calculated as follows:

$$AFL = \frac{\text{Length of the silk reeled in metres} \times \text{No.of cocoons maintained/end}}{\textit{No. of reeling cocoons}}$$

Where, No. of reeling cocoons = (No. of cocoons taken for testing - No. of converted unreelable cocoons - No. of converted carry over cocoons).

$$\text{NBFL} = \frac{\text{Length of the silk reeled in metres X No. of cocoons maintained/end}}{\text{No. of feeding ends}}$$

Higher the AFL, better is the quality of cocoons

(4) Average Non Broken Filament Length (NBFL): It is the continuous length of filament in metres that can be unwound from the individual cocoon without a break.

Where, No. of feeding ends = (No. of castings + No. of carry over cocoons converted no. of cocoons).

Higher the NBFL, better is the quality of cocoons.

(5) Single Cocoon Filament Denier (SCFD): Denier is the weight in grams of 9000 m of filament or yarn. SCFD is the denier of the cocoon filament. It is calculated as follows:

$$\text{SCFD} = \frac{\text{Weight of silk filament in grams} \times 9000}{\text{Length in metres of silk filament reeled} \times \text{No. of cocoons maintained/end}}$$

Lower the single cocoon filament denier, better is the quality of cocoons.

(6) Reelability Percentage: It is given by the following expression:

$$\text{Reelability (\%)} \quad \frac{\text{No. of reeling cocoons} \times 100}{\text{No. of feeding ends}}$$

Where,

No. of reeling cocoons = (No. of cocoons taken for testing - No. of new unreelable cocoons - No. of converted carry over cocoons).

No. of feeding ends = (No. of castings + No. of carry over cocoons - No. of converted carry over cocoons).

Higher the reelability, better is the cocoon quality.

Cocoon quality is measured by means of various parameters and it would be convenient to have a single index encompassing all these dimensions with appropriate weightages.

The absence of such a quality index is felt when one is required to represent the comprehensive quality measure. A grouping or classification of cocoons based on their quality is also not possible without the help of such a quality index. Some form or the other of quality index for cocoons exists in sericulturally advanced countries especially Japan and China, while there is no such measure in India. Though all the six parameters listed earlier are a part of the cocoon testing procedure in a laboratory, only a few of the important parameters are incorporated in commercial testing in Japan and China. This is possible because of the uniformity of the raw material quality.

Cocoon testing prior to transaction is mandatory in these two countries and cocoon price is fixed on the basis of its quality and the raw silk price that is fixed in advance for the entire year. In India cocoons are transacted without testing and price is purely based on the subjective assessment of quality made by the buyer at the time of purchase.

Cocoon Testing Procedure

Japanese Procedure

Cocoon Testing Procedure: The farmers bring the cocoons to a common place in the cocoon producing area, where the cocoons are weighed in the presence of both farmers and reelers. Cocoons for testing will be picked randomly, sealed in a bag with all relevant details and generally brought to the cocoon testing house by the reeler. The charges for cocoon testing in Japan are about 1000 yen per sample for cocoon lots of up to 1000 kg and 2000 yen for lots up to 8000 kg.

The cocoons brought by the reeler are received at the cocoon testing house, where a lot number is given for identification in both the format papers (inside and outside the cover) and the cocoons are weighed and the weight recorded. The cocoons are dried and sorted to determine defective cocoon percentage. It may be noted here that, sorting being a manual exercise, it is emphasised in Japan that the variation in sorting the defective cocoons should not be more between the persons sorting the cocoons.

Generally, it is required that the sorting persons have discussions among themselves and also verify using standard photographs, in case of doubt, so that the percentage of defective cocoons sorted does not vary between the sorters. The cocoons are test reeled adopting standard processing conditions. Cocoon filament denier, raw silk percentage, reelability, average cocoon filament length are also determined.

The Japanese grading table pertains to bivoltine hybrid cocoons and as such cannot be used for multibi cocoons. In addition, the reelability of Indian cocoons is much higher as compared to that of Japanese. Therefore, a separate grading table for Indian multibi cocoons with particular modifications in the concerned parameters is required.

Cocoon Price fixation: The objective way of price fixation in Japan has perhaps contributed to the orderly growth of sericulture in its prime days. The various dimensions incorporated in their meticulous approach are to be appreciated by everyone concerned.

Each year during February, the fanners plan for the quantity of cocoons to be produced during the next year. During April, farmers' co-operatives and reelers co-operatives hold a meeting and decide on the quantity of cocoons required for each filature and assure the farmers that the filature will buy their cocoons. Once the cocoons are produced, after collecting the sample for cocoon testing, the remaining cocoons are sent to the filatures. When the cocoons are received at the filature, 80% of the amount is paid to the farmers. In case the results are bad, and the amount paid to the farmer is more, adjustment is made in the balance payment. Generally this does not happen.

A group of farmers form a co-operative and cocoons produced by them are mixed and sold. This is possible since all the fanners of a particular area rear the same race cocoons and all the farmers in that area receive worms from the same chawki centre and variation in cocoon quality is minimum. Unless the cocoon distribution system is rationalised, the sericulture fanners cannot expect stability in the management of cocoon production. If the cocoon marketing system is stabilised, the farmers will have confidence to take up sericulture. The system provides the filatures the confidence of getting required quality/ quantity of the cocoons.

Though the cocoon transaction in Japan has been liberalised, testing of cocoons and fixing the price based on test results has been made compulsory. Though the farmers have been given the freedom to select any filature, the concept of selling the cocoons in groups to farmers' co-operatives has been made compulsory. Depending upon the market demand for raw silk in the country and outside, the demand for cocoons fluctuates. But variations in the production level of cocoons and raw silk by the farmers and reelers on the basis of market trends are not possible.

Chinese Procedure

Cocoon collection centres in China are small buildings equipped with fans, weighing scale, electronic balance and an electric cocoon drier. Green cocoons are brought by the farmers, in bamboo baskets of roughly 12-15 kg capacity. The purchase of cocoons is based on the principle of higher price for good quality cocoons, classification of cocoons on the basis of quality and fixation of price according to quality. It is reported that the policy of the government is to fix the price of cocoons on the basis of economic development and cost of cocoons required to produce one ton of raw silk.

The procedure followed for cocoon testing is as follows:

(1) Roughly 5% of the quantity from the cocoon lot is selected by picking cocoons from different parts of the lot. From this, 125g of cocoons are weighed, selecting the cocoons at random.

(2) Number of cocoons in 125g is counted. Double cocoons and bad cocoons are sorted out and corrections for total number of cocoons are made. The dry shell weight is also determined.

It is mentioned that the system does not permit sale of:

(a) Premature cocoons

(b) Lots with double and bad cocoons exceeding 20%

(c) Lots less than 25kg

Indian Procedure

Cocoons are generally not tested before transaction in the cocoon markets. The transaction is carried out by means of an open auction. The cocoon lots are displayed one beside the other and the buyers subjectively assess the quality of cocoons on display by visual and tactual means and offer their bid price. The highest bidder gets the cocoon provided the farmer agrees with the price offered. The farmer can refuse to sell his cocoons and wait for subsequent rounds of the auction. Thus, there is neither a procedure for testing the quality of cocoons nor an approach for fixing their price based on its quality and market conditions.

Therefore, there is an urgent and strong need for developing an objective method of assessing the quality of cocoons based on the parameters and an organisational mechanism to bring its implementation to cocoon transactions.

Relevance of Quality Index

Quality is an important aspect more so in the case of raw material in deciding the quality of the end product. When there are multiple measures of quality, it becomes important to represent them as a unique value or to say an index by giving appropriate weightage to the individual quality variables. It is essential to have such a measure for purposes of comparing quality of two or more lots or grading of different lots of cocoons.

Succinctly, Cocoon Quality Index can be defined as a numerical measure or an expression of the totality of its quality features.This way of representation exists in defining the quality of raw silk. Raw silk quality index is expressed in terms of grade, which ranges from 4A to B as per the international standards and 6A to E as per the Indian standards. The raw silk quality is represented by eleven parameters, which are classified as major and auxiliary ones. The major quality parameters carry equal, yet higher weightage, while the auxiliary quality parameters have differential and less weightage. The unique index representing raw silk quality is a useful means of assessing as well as comparing the totality of the features of quality of raw silk.

When there are multiple characteristics affecting quality as in the case of raw silk, a finite number (often small) of grades are set based on ranges of values of characteristics. This, they say facilitate quality-price linkages.

Ladd and Martin highlight the importance of using a single index of representing quality. They give an example of a carload of corn, which is tested for five different quality characteristics and recommends a specified-order method of describing this corn quality. However, in the case of cocoon, there exists neither a universal means of assessing its total quality nor an international standard. Each silk producing country has evolved its own procedure for assessing the quality of cocoons it produces and thereby its quality index. This index forms the basis of price fixation and hence the transaction.

Cocoon quality is determined by a number of parameters, each of which has its own relative significance on the reeling efficiency and raw silk quality. The procedures for assessment of a few of these parameters are quite involved, while it is relatively easy for a few of them. A few tests are non destructive, while a few are destructive. Thus, each of the parameters is associated with a certain extent of ease or difficulty in assessment. Further, what are available as national standards in other sericultural countries cannot be adopted as such, in India owing to many reasons. Firstly, the quality levels attained elsewhere and the existing levels in India are widely different. Secondly, the quality requirements vis-a-vis the reeling technology adopted are different in those countries. Thirdly, the infrastructure and the policy support established elsewhere facilitate detailed assessment of cocoon quality. Fourthly, no transaction takes place in other sericultural countries without testing and industry abides by the mandatory requirements. On all these counts, it appears that India cannot adopt the grading procedure followed elsewhere.

The quality of Indian cocoons are low and thus the system of grading adopted in other countries may not be capable of differentiating between the lots since all the lots are more likely to fall in to the lowest category. The infrastructure for carrying out elaborate tests on a commercial basis to deal with our entire production does not exist and there are no policy initiatives in making cocoon testing mandatory prior to transaction. Sericulture being a state subject, there is no uniform system amongst different states for transactions.

Parameters for CQI

Of the several parameters used to measure cocoon quality, shell ratio % (SR%) and defective cocoon % (DC%) have been identified as the most significant ones. They have a greater bearing on the reeling efficiency as well as yarn quality and at the same time are relatively easy to determine requiring minimum facilities, infrastructure and time.

In order to ascertain the relative importance of SR% and DC% in explaining the performance of the cocoons during reeling, laboratory data from CSTRI, in respect of a sample of around 150 lots of multibi cocoons were collected. Step wise regression analysis was carried out with all the cocoon quality parameters as independent variables and reeling performance (renditta) as the dependent variable in order to quantify the contribution of SR% and DC% in explaining the quality of cocoons.

SR% explains the maximum that it got entered first in the model, next is DC% and third is the AFL. The other parameters did not get represented in the model. The unstandardised coefficients, which provide the relative importance of the variables also indicate the significantly higher contribution of SR% and DC%. Thus, the fact that SR% and DC% together largely represent the cocoon quality is acceptable.

In the recent years quality based pricing of cocoons and raw silk has been the focus under the bivoltine sericulture technology development project supported by Japanese International Co-operation Agency (JICA). A number of short term experts from Japan who have visited Central Silk Board (CSB), Bangalore have closely studied the Indian situation and have recommended strongly in favour of introducing quality based pricing in cocoon transaction to bring in qualitative improvements in cocoon production and facilitate development of rearer-reeler linkage.

Yamamoto has pointed out that the quality of bivoltine cocoons needs to be reflected in the market price. Therefore, cocoon testing system should be introduced following two methods for measuring cocoon quality and fixation of cocoon price. He advocates measurement of good cocoon percentage and shell ratio percentage. He further adds that an electronic balance is sufficient and it takes utmost 20 minutes to carry out this work and thus is an easy method.

Yamamoto has proposed an empirical method as an illustration for fixing the cocoon price, which can reflect cocoon shell ratio and good cocoon percentage. He has expressed the need for further work to fine tune the approach so that fixation of cocoon price is made simple and ensure that the method of measuring cocoon quality is established. He admits of certain uncertainties in the approach proposed by him and suggests that the various numerical values used for trial calculation need to be examined further. He adds that since the procedure of ascertaining the correct price will become complicated if the cocoon quality is classified in to many categories, it is desirable to adopt the cocoon quality classification method, which will make the class divisions of cocoon quality as few as possible, in actual transactions. In the light of the above, there exists a felt need in the Indian silk industry to establish a cocoon grading system as a precursor for quality based price fixation.

Cocoon Transactions

To learn to understand real market situation, which is very complex, it is necessary to start by studying simple cases and the simplest of these is the perfect market. The following conditions are to be fulfilled in a perfect market as far as possible:

(a) a homogenous commodity

(b) Large numbers of buyers and sellers

(c) Perfect information among buyers and sellers on price changes and

(d) Pairing of buyer and seller by chance and not by preference.

Though, cocoons are a homogenous commodity, the quality of the lot does vary. However, the range of quality is open to inspection and thus, it can be said that they vary in a known fashion. The relationship among the price it fetches and its quality and the lot size, in addition to the information available to the buyer and seller needs to be assessed. The buyer would be interested in paying a just price for the cocoons rather than paying for the cocoons more than they are worth. The seasonally of cocoon quality and supply and volatility of prices of cocoons and raw silk do have a predominant presence in the silk industry.

A study of the performance of a reeling unit should be preceded by a study of the market situation, whereby one tries to understand the interrelationships and causative nature of the variables concerned. An understanding of the factors that influence the cocoon price and the relationship between cocoon price and raw silk price would assist in studying the performance of a reeling unit.

Prices in a competitive market equate marginal and average revenues with marginal and average costs, thus directing production and allocating resources. When one looks at the fluctuations in the cocoon price, one can observe day to day fluctuations, weekly fluctuations, monthly as well as seasonal fluctuations. Besides, on a given day, there is a range of prices offered to the different lots of cocoons. It is pertinent to explain as to what causes these fluctuations. In the absence of quality assessment prior to transaction, the subjective assessment of cocoon quality by the buyers will determine cocoon price, given the supply and demand situation. Since cocoons are perishable commodity and storage at sellers" end is not possible, market clearance occurs and supply quantity thus can represent the demand too.

If the commodity in question is not storable, then the quantity consumed cannot exceed the quantity transacted and the quantity transacted cannot exceed the quantity produced. Moreover, because wastage is non optimal for both buyers and sellers, production would never exceed sales and purchases would never exceed consumption, according to him.

The market conduct and performance in terms of its response to quality and supply position gives a better understanding of the silk industry. It is relevant to assess how much of the fluctuations in cocoon price is explained by its quality and supply quantity. In cases where the product of a sector are identified by quality, then, unless the price discriminates them, the allocation of resources might not be optimal. They say, however, the ability of the market to discriminate the prices of different qualities depends on the utility of these different quality characteristics in producing the final product, the ability of the buyers to assess the quality and competitiveness in the market.

There are three possible ways to gain unusual profits in the silk industry. One is to arbitrage between the same commodity, in different markets or between different commodities in the same market. The operations would yield unusual profits if the price differentials exceeded the costs of transportation or of processing. Market integration is thus a necessary condition for efficiency, but not a sufficient one. Attempts to set up marketing boards failed because stabilisation was a long-term solution to a short term problem, the speculative losses.

The Chinese raw silk price in India depends on the indigenous raw silk price. But, however, the causative nature of indigenous raw silk price has not been clearly established. The demand for raw silk is only a derived demand, since the demand for silk fabrics is translated into the demand for raw silk. The lead time involved in converting raw silk into a finished product being quite high and since the opportunity for storage of silk fabrics amply exists, any fluctuation in demand for silk fabrics is thus not violent and disturbing.

The demand pattern for silk fabrics in the domestic market is more or less well set, since it is identified with the festival and marriage seasons. It is reasonable to expect the demand for silk fabrics to be price elastic and income elastic. For changes in demand to be caused by the changes in price, it appears that it would be more due to the price changes from the supply side.

Reeling industry is controlled by three important input attributes viz., cocoon quality, cocoon price, and the quantity of cocoons available. Similarly, there exists three important output attributes viz., raw silk quantity, raw silk quality and raw silk price. The interrelationship among the inputs as also between the input and output forms an essential aspect of the study of reeling industry.

The raw silk traders regularly get information from the cocoon markets about the cocoon price and supply levels on a day to day basis. The loom units heavy depent on the Chinese raw silk, whose price is known to be determined by the indigenous silk. Thus, the question arises as to what decides the price of Indian raw silk. How much of its fluctuation is dictated by the supply side is a major question to be answered.

Farmers who bring big lots of cocoons of say around 300 kg to the cocoon market, have a tendency to divide the lot into smaller lots of say 100 kg each. At the same time, too small a lot would mean more heterogeneity due to mixing of several small lots at the reeler's end. There is no clear indication as to what constitutes the more preferable lot size from the point of view of buyers. There has not been much information about what the average lot size of the cocoons has been and the price it commands in relation to its quality.

Market Dynamics

This examines the various research methods and approaches used to study the dynamics of markets particularly of the agro industries. Kulshreshtha and Wilson developed an econometric model of the Canadian cattle sector, which includes the prevailing demand, supply and price relationships. The demand for beef at the retail level was found to be positively related to consumer incomes and the price of pork and negatively to its own price. Among these variables, the retail price had the greatest impact. Price and income elasticities of demand and supply were also calculated.

Grossman analyses the nature of quantities in market disequilibrium. He mentions that quantity supplied is analogous to quantity demanded. The quantity supplied which is the quantity sold, would be consistent with the maximisation of seller objectives, given the existing price and any other constraints on seller behaviour. If exchange is voluntary, the quantity supplied is also the maximum quantity, which sellers would sell at the existing price.

Roy and Ireland have made an econometric analysis of sorghum market. The objective of the study was to identify and estimate the major structural relationships, which influence annual sorghum prices in the U.S domestic market. The postulates regarding the major causal relations in the sorghum sector were generally corroborated by the results of the model. In recent years, the free market forces of supply and demand in the sorghum sector have played a relatively strong role in the price-determining process, which in earlier years was largely dependent on the price support program and credit support policy.

Pasour and Bullock discuss the problems posed by uncertainty and information as a scarce resource in analysing the efficiency of existing firms and markets. The limitations of efficiency measures, which exclude uncertainty and information are analysed and the implications for research are briefly discussed. Aspects of efficiency norms, optimal plant size, market imperfections and inefficiency are discussed.

Stifel analysed Thai rubber marketing in a framework, which recognises the multiple dimensions of the relationships constituting a market system. The analysis traces a causal sequence from market structure to market conduct to market performance. In an overview of the rubber marketing system, the author mentions that the problem of rough estimation of the quality of a certain type of rubber is acerbated by the low quality and high variability of Thai production. He says that the market conduct in this industry has accommodated the poor and variable quality of production rather than encouraging improvement and standardisation, with the result that Thai rubber is currently produced and processed essentially the same way as it has been for many decades. This reflects largely, the state of affairs in Indian silk industry.

Ladd and Martin have applied the product characteristics approach to production inputs and summarised under two themes. This approach views a product as a collection of characteristics. Product heterogeneity may arise due to product differentiation, quality and grades and standards. The first theme is that the price of a purchased input equals the sum of the money values of the inpuf s characteristics to the purchaser. The second theme is that the demand for an input is affected by the input's characteristics. They cite the example of a competitive firm producing goods that are inputs to other firms. They say, because the firm is in a competitive industry, it cannot affect the price it receives by varying the amount it produces. It can however affect the price it receives by varying the 'quality' of the output. What quality of product should it produce to obtain maximum profit? This is highly relevant in the context of silk reeling industry.

Meilke has attempted to determine the supply response of U.S pork producers to changing hog and feed prices over time and to see if the hog-corn ratio is an adequate predictor of hog supply when the prices of both hogs and feed are volatile. The analysis supports the hypothesis that the hog-corn price ratio may provide a good explanation of hog supply when corn prices are constant but not when they are fluctuating.

Traill et al developed an alternative method for estimating irreversible supply functions by segmenting the price series. They have suggested modification for Wolffram's method for measuring asymmetric response of supply function. They say that there is a strong case for estimating non reversible supply function for many commodities. Where there is a ready substitution in supply, commodities are likely to share fixed assets and the response to a change in relative prices is merely reallocation of resources towards the more profitable enterprises.

Leuthold and Hartman have constructed an economic forecasting model to serve as a norm against which to test futures market forward-pricing abilities. A semistrong form test of efficiency examines whether or not the prices in a market reflect all publicly available information. The results demonstrate that the live-hog futures market cannot be relied upon to reflect accurately and consistently subsequent cash prices and hence it is inefficient. Active trading by commercial firms may aid in rational price formation. If lack of financial capital causes the market to react slowly to information, either reducing margin requirements or contracting size, may encourage additional traders to the market, thereby improving forward-pricing efficiency. If presence of more number of traders can improve forward-pricing efficiency, there might be something optimistic in the context of the Indian silk industry.

Beenstock and Bhansali have built a dynamic structural model for agricultural prices by two separate approaches of econometric model and stationary time series. The results emphasise the limitations of using linear time series models, which are essentially developed for analysing integrated Gaussian process, on more complicated non-Gaussian and non-stationary time series, such as cocoa series.

Buccola has developed an approach to the analysis of feeder cattle price differentials. The results suggest that break-even analysis can be used successfully to clarify the understanding of market price differentials between different lots of feeder cattle. The author observes that the same variable that affect feeder price levels themselves also affect the price differentials observed between different classes of feeder cattle. He makes an important observation that break-even prices may not be used reliably to estimate short-run buyer or seller reservation prices. At times, the seller may be willing to accept a feeder sale price below that required to cover all sunk costs if he perceives that this price is what most buyers will be willing to pay. Alternatively, short-run seller reservation prices could more than cover total production costs in years of strong buyer demand. This is particularly true of raw silk market.

Ziemer and White have analysed the disequilibrium price behaviour in the U.S fed beef sector. Although tests of equilibrium versus disequilibrium hypotheses are not well developed, results indicate the possible presence of disequilibrium prices in the market for fed beef sector. The authors note that in situations where proposed agricultural policy might severely affect commodity markets, a disequilibrium model may provide policy makers and analysts with useful information concerning the speed with which normal competition behaviour will resume.

Bessler and Brandt discuss various methods for testing causality between two time series variables. These methods are then applied to movements in livestock prices, livestock quantities and income. Finally, the implications of the results for modeling and forecasting are discussed. The methods discussed in the paper include, Piercers two step procedure where each series is first prefiltered using a separate ARJMA model, Direct Grangner test employing ordinary least squares regressions, Sims procedure involving a common prefilter for ARIMA, Geweke's procedure and finally bivariate transfer functions due to Ashley et al. The author however chose to apply the direct Granger test to cattle and hog markets. The study provides specific evidence on leads and lags of several variables in the hog and cattle markets. This evidence agrees generally with a priori beliefs. The present analysis is performed without use of prefilters and a priori notions on leads and lags. They suggest that further applications of this procedure may yield fruitful results in other markets.

Ravallion offers an approach of testing agricultural market integration along the lines of a dynamic model with instantaneous market integration and the short-run dynamic adjustment process. He illustrates maricet integration using data on the interregional price differentials for rice in Bangladesh. Two markets are said to be spatially integrated, if trade takes place between the two, price in importing region equals price in the exporting region plus the unit transport cost incurred by moving between the two. Market integration is by no means sufficient for the pareto optimality of a competitive equilibrium. Measurement of market integration can be viewed as basic data for an understanding of how specific markets work. The existence of significant localised market characteristics also indicates that arbitrage is imperfect in eliminating price differentials.

Willet and French present a dynamic econometric model of the U.S beekeeping industry for policy analysis and economic projections. Data from 1952 to 1984 were used to estimate the model by three-stage least squares. The model indicates that, when the federal price support for honey exceeded the market price, the federal program had a significant impact on several sectors of the industry. A comparison of beekeeper revenue, consumer expenditures, and federal government expenditures suggest that the honey support program was an effective means of supporting honey prices.

Weersink and Tauer examined the direction of causality between dairy farm size and productivity for each state in the U.S, using multivariate Granger-causality tests. The results only partially supported the view that technological change has caused increases in dairy farm size. The direction of causality appears to be more from size to productivity. Changes in productivity and average herd size appear driven by price changes. The authors bring out very forcefully his observation that changes in both productivity and average herd size are largely brought about by price changes. An important policy implication is therefore that agricultural research in dairy farming may not lead to larger dairy farms. Instead, economic factors (prices of inputs and output) and other forces leading to structural changes may lead to larger dairy farms which are then in a position to adopt the new technology. The major determinant of changes in dairy farm structure may be dairy price policy. One could draw a parallel from this research in the context of the growth of Indian silk industry. These findings perhaps provide indicators for suitable policy measures for bringing in any structural changes in silk reeling sector in India.

Goodwin and Schroeder empirically evaluate spatial linkages in regional cattle markets using cointegration tests of regional price series. Several markets were not integrated over the period from 1980 through 1987. However, significant increase in cointegration of several regional livestock markets was observed through the 1980s. The increased cointegration parallels significant structural changes in the livestock industry. A formal analysis of market characteristics reveals that distances between markets, industry concentration ratios, market volumes and market types have significant influences on cointegration relationships between markets. Goodwin observes that market volume also may influence trade activity across market areas. Low-volume markets have a higher potential for exhibiting unwarranted price behaviour.

Analytical Study on Cocoon Markets

An analytical study conducted in Karnataka by B.V Vasumathi on silk reeling operations reveals more about the marketing situation of cocoon. The data collected for the analysis comprised cocoon price, cocoon supply quantity, Cocoon Quality Index in respect of five major cocoon markets in Karnataka, and raw silk price from Bangalore silk exchange. These five cocoon markets account for around 70% of the total cocoons transacted in the state, while Bangalore silk exchange is the biggest in the state, accounting for almost 85% of the raw silk transacted in the silk exchanges in the state. Daily data for a period of two years from 1st July 1996 to 30th June 1998 were collected. Data in respect of around 100 lots of cocoons, each day were sampled. Day wise weighted average was computed in the case of cocoon price and CQI and used for the first two objectives, while lot wise data was used for the third objective.

The analyses for the first two sub-objectives were carried out in two stages. In the first stage, the state average data aggregated over all the five markets were used. Later, similar analysis was conducted using data in respect of only Ramanagaram market. Ramanagaram market is said to be the largest in Asia and considered a price leader, accounting for almost 50% of the cocoons transacted in the state. Market wise analysis as well as overall analysis was carried out using lot wise data for the third objective.

Since the data were known to be highly autocorrelated and non-stationary, the normal techniques of regression analysis could not be applied. Time series analysis viz. Auto Regressive Integrated Moving Average (ARIMA) was applied to study the interrelationships. The time series and dynamic models are used in three important areas of application:

(1) forecasting of future values of a time series from current and past values

(2) determination of the transfer function of a system i.e., determination of a dynamic input-output model that can show the effect on the output of a system of any given series of input and

(3) design of simple feed forward and feedback control schemes by means of which potential deviations of the system output from a desired target may be compensated as far as possible.

Determination of the transfer function of the system is the objective of using the ARIMA in this case. Dynamic regression models or transfer function models are used to study the existence of any interaction between the series. An important type of dynamic relationship between a continuous input and a continuous output is that in which the deviation of input variable (X) and of output variable (Y) from appropriate mean values are related by a linear differential equation. Ordinary regression with serially correlated time series is unreliable.

Most time series have some trend either up or down, and any two trending time series will correlate simply because of the trends, regardless of whether they are causally related or not. Thus, in order to regress one time series with another, estimates of the linear relationship are required apart from accidental similarities resulting from autocorrelation. The autoregression procedure which estimates true regression coefficients from time series with first-order auto correlated errors was applied to study the dependence of cocoon price on CQI and supply. Maximum likelihood regression with autocorrelated errors was used. It is cited that, this algorithm gives the best possible estimates.

ARIMA Model Building

SPSS was used for the statistical analysis. The first step of the iterative approach for ARIMA model building is model identification. The values of p, d and q i.e., the difference, autoregression and moving average operators respectively were identified as per the procedure of Box & Jenkins.

The missing data were filled using the criteria of mean of two adjacent values (linear interpolation method). The presence of seasonal process is determined, if ACF, PACF or both show significant values at lags that are multiples of the seasonal period. In this case, significant values were observed at lags, which were multiples of 7 and since daily data was used, a seasonal period of a week i.e., 7 days was used in the model. The principal problem in identifying seasonal ARIMA models is the complexity of the ACF and PACF plots. These plots arise from the combination of the seasonal and nonseasonal ARIMA processes with random noise and are rarely explicitly clear. SPSS Manual recommends, to initially identify some model, estimate the coefficients, obtain a residual series and then inspect the ACF and PACF of the residuals for clues about components we need to add to the tentative model. The ARIMA cycle of identification, estimation and diagnosis takes longer when seasonal processes are present. A multiplicative model was fit, wherein the seasonal decomposition procedure treats the series as the product of the seasonal, trend and cycle components.

First, it is required to identify the seasonal model and later identify non-seasonal model for estimating the complete model. The ACF plots from the seasonally differenced series were inspected and the tentative model was taken as (1,1,1). The parameters of this tentative seasonal model were estimated. Constant was not included in the model since the mean seasonal difference should be about zero. By estimating the seasonal model, we obtain a residual series free of seasonal effects. If the identification of the seasonal model was correct, these residuals show the non-seasonal portion of the model, else they will continue to show autocorrelations at the seasonal lags. The ACF plot suggested a good identification of the seasonal model. In the next step the non-seasonal components of the model were identified. The nonseasonal model was decided based on the ACF and PACF plot of the residuals from the seasonal ARIMA model, as (1,1,1). The tentative model incorporating both seasonal and non-seasonal effects is thus (1,1,1) (1,1,1) 7. The initial and final parameters, along with iteration summary were generated.

The next step is diagnosing whether the tentative model fits the data or not. Here, the residuals are checked to see if they were white noise. The ACF for the ARIMA residuals were plotted and none of the autocorrelations exceeded the confidence limits around zero. The Box-Ljung statistics did not show significance at any lag and thus the null hypothesis that the residuals are white noise cannot be rejected. The Box-Ljung statistics are given in Table 1.

Table 1. Test of Significance of the Residuals of the Fitted ARIMA Models

Source	Variables	Box-Ljung Statistic	Prob.
State Av.	Cocoon Price	13.530	0.561
	CQI	15.000	0.451
	Cocoon Supply Quantity	14.447	0.492
Ramanagaram Cocoon Market	Cocoon price	13.319	0.578
	CQI	22.230	0.102
	Cocoon Supply Quantity	13.566	0.559
Banglore Silk Exchange	Raw Silk Price	8.229	0.914

The cross correlation functions (CCF) of the series were observed. The CCF showed the correlation between two series at the same time and also with each series leading by one or more lags. By inspecting the CCF between two series, one can see the lag at which they are most highly correlated.

Cocoon Price Vs CQI and Supply Quantity

In the case of state average data, supply leads the cocoon price by a lag of one day. There is a significant correlation at lag one between the two series. However, there is an instantaneous causality between cocoon price and CQI. They show significant correlation at lag zero. The lag correlation between cocoon price and supply as also CQI was found to be significant at lag zero in the case of Ramanagaram cocoon market data is zero. Thus maximum likelihood regression was run with cocoon price as the dependent variable and CQI and supply lagged by one day as independent variables in the case of state average. Similarly, the regression was run for Ramanagaram cocoon market data, with lag zero.

The estimated parameters and the goodness-of-fit of the first order ARIMA model are as given below:

(a) State Average:

$$W_t = 0.6192\, W_M - 0.00275\, W_{t\_7} - 0.87707\, a_M + 0.996\, a_{t\_7} - 2.068\, CQI_t - 0.0006 S_t + a_t$$

(11.716) (14.281) (25.944) (14.289) (6.1903) (3.86)

$R2 = 0.748^{**}$

Figures in the parenthesis indicate t values.

** Significant at 1% level

Where W_t = (1-B) (1-B7) P_t is the working series

a_t = Residuals

S_t = Cocoon Supply Quantity

CQI_t = Cocoon Quality Index

Wt =0.6398Wt.1 + 0.0253 Wt.7 - 0.8461arl- 0.9857 at.7 - 5.918 CQIt - 0.000135 St + at

(9.78) (0.63) (17.90) (44.52) (13.17) (3.569)

$R2 = 0.835^{**}$

Figures in the parenthesis indicate t values.

** Significant at 1% level

Where W_t = (1-B) (1-B7) Pt is the working series

a_t = Residuals

S_t = Cocoon Supply Quantity

CQI_t = Cocoon Quality Index

It may be observed that around 75% of the variation in cocoon price in the case of state average and around 84% in the case of Ramanagaram cocoon market, are predicted by the model.

Cocoon Price Vs Raw Silk Price

Similarly, CCF was computed between cocoon price and raw silk price. It may be observed that in the case of state average, cocoon price leads the raw silk price by a lag of 7 days. There is significant correlation between the two at la<* 7 However, in the case of Ramanagaram cocoon market data, this lag between the cocoon price and raw silk price reduced to 4 days.

Maximum likelihood regression was also run with raw silk price as dependent variable and cocoon price lagged by 7 days as independent variable for state average data. Similarly, the regression was run for Ramanagaram cocoon market data with a las of 4 days.

The estimated parameters and the goodness-of-fit of the first order AREVIA model are given below:

(a) State Average:

$$W_t - 0.0813\ W_{t-1} + 0.0648\ W_{t-7} - 0.824a_{t-1} - 0\text{-}9998a_{t-7} + 0.824CP_t$$

(1.73) (1.63) (30.05) (1.015) (2.415)

$R2 = 0.463^{**}$

Figures in the parenthesis indicate t values.

** Significant at .000

Where W_t = (1-B) (1-B^7) P_t is the working series

A_t = residuals

CP = Cocoon Price

(b) Ramanagaram Market:

Wt = 0.07709 Wt-1 + 0.07 Wt-7 - 0.835at-1 - 0-9996at-7 + 1.07CPt

(1.73) (1.63) (30.05) (1.015) (2.415)

R2 = 0.457**

Figures in the parenthesis indicate t values.

** Significant at 1% level

Where W_t = (1-B) (1-B^7) Pt is the working series

a_t = residuals

CP = Cocoon Price

The model is able to predict around 46% of the variation in raw silk price in the case of both state average and Ramanagaram cocoon market.

Cocoon Price and Raw Silk Price

ARIMA studies have established the dependence of raw silk price on the cocoon price i.e., the cocoon prices lead the raw silk price by 7 days. It would appear, rawsilk market uses information from the cocoon market, while the cocoon market is not really making use of the complementary information from the other market. The leadlag relationship between the prices of the two markets will be providing for profits through arbitrage. This affects pricing efficiency. It is said that significant crosscorrelations at non-zero lags indicate price leadership.

The lack of instantaneous causality between cocoon price and raw silk price implies the existence of some barrier between the two price series. This can be explained by the processing time involved in converting cocoons to raw silk, which generally ranges from 3 to 7 days. The yarn price at period Y and cocoon price at period (t-T) are significantly related.

However, the lag got reduced to four days in the case of Ramanagaram market. The reduction in the lag period suggests the importance of price indicators from Ramanagaram market, which has been revered by the industry as a leader, reflecting the market dynamics more thoroughly. Thus it could be construed that the cocoon price indicators from the other markets are sensed over a period of 4 to 7 days.

The chain of dependence can thus be represented as:

Cocoon supply → Cocoon Price → Indigenous Raw silk price → Chinese raw silk price

The relationship between cocoon price and cocoon quality index at lag zero is understandable since logically the cocoon price can only depend on the quality of the cocoon on that day and not on those for the previous days or at a later date. The lag of one day between supply and cocoon price, in the case of state average needs to be reasoned out.

When a reeler comes to a particular cocoon market, the supply position at that market can only influence him and what may be available with him is the information on the supply position in the other markets on the previous day but not for that day. Though supply in the five markets are well correlated, the rate of change in supply across the markets may not be the same and relative change in cocoon price in proportion to the supply, thus, will neither be the same nor instantaneous in relation to the overall supply. This is supported by the outcome of the analysis carried out for only Ramanagaram market. Here the lag between supply and cocoon price is zero and there is an instantaneous transmission of supply information onto the cocoon price.

Product price is a given datum to the individual firm, but at the same time, the supply plans of all competitive producers as a group, are a basic determinant of product price. Though each firm supplies a negligible fraction of the total, supply curves of all the many firms in the industry constitute the industry supply curve. This curve does have an important bearing upon price. Even in the absence of a system to facilitate quality based price fixation, the significant dependence of cocoon price on cocoon quality index reflects the ability of the buyers in making a subjective assessment of cocoon quality.

Simple correlation studies have established that the heterogeneity of the cocoon lots in terms of quality and lot size on a given day is reflected in the spread of the cocoon price. Seasonally in cocoon quality is a known factor. The fact that the buyers are able to strike an equilibrium with the prevailing quality levels is something to be taken note of. Overall quality levels are reflected in the average price levels of the cocoons. Cocoon quality is a very critical factor in deciding the output and in turn the economics of the reeling activity. Obviously, the reelers would not be prepared to pay more for the cocoons than what it is worth. Had the influence of the demand sector been significant, the equilibrium of price and quality could not have been so consistent over time, that the reelers would face situations where the price they pay for the cocoons is much higher than what it deserves in terms of quality.

Regional silk exchanges may be linked through an oligopsonistic interdependence where reeling units compete within a limited service area. This situation may result in market linkages that follow a non competitive basing-point pricing which is maintained through an organised oligopoly arrangement. This leads to the establishment of a base price at a central·location and the delivered price to any buyer location is the base price plus transportation costs. This is particularly true of silk markets. Hardly 20% of the silk produced in the state get transacted in the recognised markets and the remaining silk gets transacted on a private basis. The traders and weavers are concentrated in distinct locations. Bangalore silk exchange is the central location where the basing-point pricing is established, since the feelers from here about the price movements go up to Varanasi on an hourly basis, not to mention about the smaller silk exchanges in other parts of the state.

Concentration is typical of reeling units, weaving units as also the trading units. With increased concentration of units across several markets, cointegration of these markets could increase, especially if the highly concentrated firms compete in the same market regions. Carlton has shown that price stability increases with increasing concentration in several industrial markets. The increased concentration, he says, may facilitate collusive pricing behaviour resulting in discriminatory basingpoint pricing practices.

For a pair of prices from speculative markets, cointegration implies a lack of efficiency because prices should incorporate all available information. However, this result requires instantaneous arbitrage, which is necessarily precluded by spatial linkages that include delivery lag. With such linkages, a spatial dimension to efficiency will require that prices do not wander arbitrarily far apart in the long run. In this light, cointegration may imply a form of spatial efficiency for regional market linkages.

Researches had indicated interdependence of cocoon price and silk price, while the present analysis indicates a causative relationship and the supply sector is influencing the pricing of the end product or to say the demand side. Looking back at the change that has taken place in the industry over the last few years, the ease of availability of Chinese silk has enhanced. Huge quantities of smuggled silk are available in the market, which is not officially kept track of. The dependence of power loom sector on Chinese silk has increased, thereby making them less dependent on indigenous silk. Informal discussions with the traders in Varanasi silk market, which is the focal point of Chinese silk transactions, indicated that Chinese silk consumption has increased from around 1400 tons during 1990-91 to around 2000 tons during 1999-2000. Thus, the fluctuations in the demand for silk fabrics in either the domestic market or the export market is not passed on to indigenous raw silk market so as to cause fluctuations in demand for raw silk that would in turn reflect on the cocoon price.

The indigenous raw silk is largely consumed by the handloom sector and partly by the power loom sector as weft, provided the price is marginally lower than that of Chinese raw silk. The handloom sector cannot quickly respond to the market demand as to translate them instantaneously as demand for raw silk. The power loom sector's dependence on indigenous raw silk is neither total nor critical. These factors might have weakened the influence of the demand side market on indigenous raw silk price that the interdependence of price relationship shifted to a state of causality.

Generally, a basic economic assumption is that at the market price, quantity transacted is equal to the minimum of demand and supply, or in other words, the short side of the market dictates the level of exchange. In case of silk industry, it is clearly the cocoons, which is in short supply. Though conventional wisdom attributed the growth of exports almost exclusively to the demand side, the case of silk is different. The growth of silk industry according to the result of an econometric model, is evenly shared between demand and supply side. Federico says that it is difficult to believe that silk exports alone could have caused development and probably the causation has to be reversed from development to silk exports instead of the other way round. This again, reinforces the predominance of supply sector in the silk industry.

15

Issues in Silk Reeling

The fundamental factor that maintained the relative profitability of sericulture was not so much a favourable price for cocoons but was the rapid and sustained progress of technology in this particular sector. The expansion of raw silk industry and accordingly, the improvement of its competitive power in international trade depended heavily upon technological progress in sericulture. In fact, it is an outstanding example of a situation in which the semi-factoral terms of trade moved much more favourably than the commodity or barter terms of trade. The governments recognised the importance of quality improvement and enacted a law to curb the production of raw silk of inferior quality. Efficiency in the silk industry was enhanced by the introduction of more economical and better controlled methods of producing silkworm eggs.

For its producers and for the great bulk of India's people, silk is of use only as a commodity to exchange with the wealthy. The danger of the increased application of technology to development in this instance as more generally, is that even this utility becomes more and more constrained as the process of production becomes increasingly inaccessible to any but the already rich. How sericulture can continue to flourish in a competitive world and yet provide employment, income and scope for enterprise to as broad a section of the population as it hitherto did, is a big question of the current period of sericulture development.

Irregularity and variation in production, involvement with subsistence fanning, lack of any documentation or accounting in most of the industry and hostility to financial enquiry among the few who may undertake some accounting for their own benefit; all these combine to make it extremely difficult to obtain the data necessary for financial evaluation.

Naik .G and Singh attempted to build a model of the silk industry considering the aspects of mulberry sector, cocoon sector, raw silk sector and silk fabrics sector. The parameters of cost and quantity in aggregation have been considered without delving into the aspects of quality and its nuances existing in the present scenario. This study aimed at understanding and quantifying the relationship between important variables of the various sub-sectors of Indian silk industry by developing an appropriate econometric model and using it for policy simulations.

Aspects of forecasting the industry scenario for the future period, effect of changes in the import price of raw silk, effect of changes in the export price of fabrics have been studied. It is admitted that a clear analysis of the effects of increased imports on production of high quality silk in the country is beyond the scope of the model. Since production of good quality silk has been a problem area coming in the way of progress, a simulation model incorporating the aspects of quality of cocoons and raw silk, establishing their inter-relationship may yield useful insights.

The changing pattern of silk exports from India and quantification of import demand elasticities for silk fabrics in the world market have been studied. The silk fabric exports as a variable dependent on the total production of silk yarn in the country, the GDP of importing countries at constant price, the income elasticity and price elasticity in various importing countries have been studied. The non-traditional items like Ready Made Garments represents the type of products that has a high growth potential. For countries with low income elasticity and price elasticity, the strategy should concentrate on value addition, the study has indicated. The markov analysis showed that dress materials will continue to be the major item among the silk goods exported from India and there has been a changing nature of export demand in favour of high quality non-traditional goods.

Report on the international conference on silk has highlighted the subtle problems coming in the way of orderly development of the silk industry as also the measures of overcoming them, which are very much relevant in Indian scenario,

(i) *Prices*: Raw silk needs price stability taking into account equitable cost and inflation factors for producers and trade. Both must have a sufficient remuneration to compensate for risks taken and functions exercised,

(ii) *Orderly Marketing and Communication*: The imbalanced pricing structure of raw materials, semi-finished and finished silk products needs to be remedied. The confusion of fashion makers and consumers as to what type of merits they can expect from the different silk products in the market is to be put to an end by an adequate communication,

(iii) *Balancing Supply and Demand*: A careful Supply and Demand management in accordance with worldwide production and demand trends is recommended and one follows with interest the respective unification efforts undertaken in China. (In China a plan for further reduction of reeling capacities to balance the supply and demand is in force),

(iv) *Quality*: There's an urgent appeal by the consuming countries to the producing countries to undertake even more efforts to improve the quality of raw silk. Close focus on this problem through contacts and relevant dialogues between producers and users is advocated,

(v) *Reciprocity:* With the advancement of all silk products from producing countries to the user's countries, the latter express the wish of reciprocity as a desired goal for the mutual benefit of both sides.

Issues Related to Cocoons

There are various issues pertaining to Indian silk industry. The volume of cocoons available varies from one season to another by a factor of ten. The inevitable implication is that only one tenth of the reeling can occur at seasons of scarcity as compared with seasons of plenty. The operation of most reeling businesses cannot therefore be continuous, and whether to continue or to halt is in fact for most reelers a matter of almost daily decision. The decision has to be taken in the light of the prices of cocoons and silk, together determining whether a profit from current reeling is likely or not.

Issues Related to Reeling

The equipment of silk industry was notably simpler and less capital intensive than that of other industries. This feature of technology has three very important consequences:

(a) The value added is very low.

(b) Reeling needs remarkable manual skill and a lot of attention.

(c) Scale and scope economies are negligible i.e., the equipment is totally specific and perfectly divisible. Plant size and productivity per basin had no correlation.

These three characteristics created a distinctive economic environment - highly competitive, very risky and inhabited by many small firms.

All over the world the industrialists complained about excessive competition. Arguably, the competition was a consequence of the absence of scale and scope economies, which removed the most pressing motive for concentration and increase in the size of firms. Two pre-requisites that existed were, that no firm had a large share of total sales and the barriers to entry were low. The great ease for any one to speculate with cocoons and to play the role of a reeler, even by producing low-quality and not profitable silk, the high turn-over and the low barriers to entry can be considered as structural features of the silk industry.

A successful firm needs good labour management and the best trading skills. The labour management is essential because, careless work could reduce the productivity and/or worsen the quality of silk and therefore lower its price. Trading skill (shrewdness and common sense in trading) is extremely important because profits were so much dependent on the result of speculation. The secret of a successful manager is to be able to discover the best buy for cocoons and to sell the silk at the right moment (i.e. to forecast well the silk price). Business in reeling is thus an art. not a science, he opines.

The profits in silk reeling arise from the management of this business. The principal part of the profits of this business is not industrial profits, but commercial ones, and is obtained with the fluctuation of the material cost and the product price. The profits of this business therefore arise from the process of non-industrial and commercial transactions.

Hishitani has looked into the possibilities for improvement of the system of business of silk reeling industry. One of the weak points of the management system of this business is the low value added per capita of the labourer. The second weak point, he says, is the shortage of owned capital to buy the material. The level of investment on machine and equipment in this business is quite low, but it is very difficult to invest more on machine and equipment to increase labour productivity due to the shortage of owned capital. The burden of this further interest becomes vital to the safety of the finance of this business. He says that the problem of this business should be solved as the system, especially to buy the material as cheaply as possible. He concludes that thus business of silk reeling should study the area of buying the cocoons as a problem of the system.

Kiyokawa narrates the experience of the Tomioka model filature in transplantation and adaptation of the European factory system in Japan. This filature was designed by the government for the purpose of improving the quality of Japanese raw silk since silk was the most important export good in the early period. The technology was imported in total from France. The marketing policy of the filature was also similar to the policy broadly adopted in France. Cocoons were strictly selected after machine-drying, and fine raw silk of high quality was carefully produced at the cost of reeling efficiency. Although this "quality-first" principle as well resulted in a high proportion of waste silk particularly in the case of poor Japanese cocoons, Tomioka raw silk obtained a high evaluation in the international front. These conditions were too idealistic for the pre-industrial society of Japan. Accordingly, the financial situation of the filature had, as a natural course of events, shown successive losses.

Referring to the economic adaptations in silk reeling industry, Kiyokawa describes the various modifications in technology, marketing policy and the institution that were introduced. Of all things, the fact that the quality of Japanese cocoons were much inferior to those of Europe was the starting point of adaptations in technology and marketing policy. For instance, it was realised that it was not an efficient policy to produce 12 denier silk from Japanese cocoons. The production of fine deniers from poor quality cocoons is accompanied by a high proportion of waste silk. The main product in the Japanese silk reeling industry, therefore, was rapidly shifted to a thicker size raw silk. These products were welcomed in particular in the U.S market. A switchover from quality-first to quantity-first principle immediately encountered a serious problem, namely, the insufficient supply of cocoons. This difficulty was gradually solved in the first decade of the century by rapid development of bivoltine silkworm varieties.

The Tornioka model filature directly contributed a great deal in diffusing the knowledge of the new filature system. The rapid diffusion of the system was indirectly supported by the swift social adjustments due to active entrepreneurship and flexible social value system. The growth and development of Japanese silk industry from low quality levels to high quality levels and the strategic approach adopted especially in dealing with inferior quality cocoons and the introduction of factory system of reeling with good management would perhaps provide the much needed direction to Indian silk industry.

Kiyokawa studies the quality of labour in silk reeling filatures in Japan. More typical features of disciplined labour and the Japanese factory system are observable in the silk reeling filatures. The technology of silk reeling industry was a kind of appropriate technology. However, for the technology to be appropriate, the unfamiliar factory system of the west had to be more appropriately adapted to the real situation of the economy of Japan. Typically, the labour intensive technology demanded efficient labour under strict labour management. Raw silk is a quality sensitive product. Quality control, therefore necessitated strict product inspection and process control even under the less machine dependent technology or perhaps all the more. Thus labour and factory management had to be arranged to satisfy the required level of product quality.

When Japan tried to explore the American market for lower quality raw silk for thicker fabrics woven on power looms, it needed to modify the transferred European technologies into a technology that suited Japanese market conditions. Thus, appropriate technology was developed through uniting selectively, parts of the Japanese indigenous technology and European traditional and modern technologies. To be competitive, it was not sufficient to have chosen to adjust production technology so as to produce low quality silk. It was also necessary to supply low quality product at a relatively lower price. In case of raw silk production, the main source of cost savings was in the formation of efficient and cheap labour, since the production was labour intensive with delicate finger work. The role of labour management was thus crucially important in the factory system. Improvement in product quality became inevitable in the 192os owing to the demand shift in the American market.

While in terms of technology, silk reeling technology is scale neutral, the average firm size had gradually increased and large scale firms had a decisive influence on the industry as a whale.In filature, labour productivity or skill was measured according to three different aspects, viz., reeled quantity, effective reeling rate of raw materials and the quality of reeled products. Measurement of quality was the most difficult, but the most important in relation to product price. He highlights the importance of supervisor's job in a silk reeling unit in maintaining the quality. The gradual shift of production policy from a low quality low cost to a high quality high cost type was inevitable for Japanese silk reeling industry'. This could be realised directly by intensifying quality control and production process control and indirectly with quality improvement of raw materials and technological innovations in the industry in the long-run.

The Japanese experience suggests that a rational labour management supported by sufficient technological knowledge contributed most to extracting the potentiality of unskilled rural female workers. The provision of technical education for managerial staff from early days was a key variable in gradually modernising labour management, in parallel with the successive introduction of technological innovations. What was more important as a background factor in the case of Japan was keen competition among firms, which encouraged labour management to function effectively.

Raghavendra et al studied the economics of Indian silk industry. The net revenue of a reeling enterprise was calculated based on primary data collected at a point of time. The study indicated more or less negative net revenue in many instances. The charka and filature sectors were dealt separately in the study. The spread sheet based simulation of the profitability in reeling operations discussed, did not consider the temporal aspects related to seasonality, while the analysis of the trend in cocoons and raw silk indicated a high degree of seasonality.

Technology Integration

Ono has studied the characteristics of technical progress in Japanese silk industry. He evaluates the importance of small innovations, points out two typical cases of borrowed technology and lastly, measures the rate of technical progress in machine filatures. The discussions and analysis are relevant in the Indian context as it contains a few lessons for Indian silk industry in the adoption of technology. He states that entry into silk industry can be assumed to be almost free. His analysis of the data of the survey of silk reeling factories in all Japan, points out to the entry and exit phenomenon among large scale as well as small scale factories. This, he says, enables one to characterise the silk reeling industry as being fairly competitive. Competition is, on the one hand, often accompanied with waste of capital equipments due to bankruptcy. But on the other hand, it should be appraised as a motivational factor, he observes. Competition among silk manufacturers in Japan induced their desperate efforts for survival and gave rise to a great number of innovations.

Ono says that it is widely accepted that borrowed technology is one of the primary factors, which ensures a high rate of industrial growth in developing countries. He defines two types of borrowed technology. If the technology presupposes an advanced knowledge of natural science far beyond that of the developing country, and development of a particular method of production requires a great amount of money, it will be virtually impossible for the developing country to make its own equipment reflecting domestic factor prices. The best and the only way the benefit of advanced technology can be shared in this case is to import the completed equipment in an original form, even though it is incongruous with the factor prices. The rigidity of factor proportions and their incongruity with domestic factor prices are the distinctive features of the first type of borrowed technology.

In the second type, a developing country is able to approach the same body of technological knowledge that an advanced country has, because the technical gap is narrow between the two countries. The developing country develops by itself the other part of production function, by applying the new technological knowledge acquired from foreign made equipment. A less capital intensive method of production is chosen in the developing country and thus a higher rate of profit is realised than in the case of the first type of borrowed technology.

In the estimation of the rate of technical progress in silk industry, Ono has worked out the elasticities of silk output with respect to capital, labour and materials. Small innovations, if they take place frequently, will have remarkable effect on the long run trend of labour productivity. These help the developing countries to cultivate their own originalities, since small devices are much easier to make than spectacular innovations. Higher labour productivity allowed Italian firms to pay their workers more than the Japanese ones and to have roughly the same labour cost per unit of output. The secret of low cost of Japanese silk lay not in the wage cost but in the cheap price of cocoons which, in turn, was due to the low opportunity-cost of labour in the country side.

Federico says that the solutions by way of low capital-intensive technology were surely sensible in the short and medium term, but in the long run they were not compatible with the desired productivity growth. He discusses the response of traditional industry to new technology and the description of the process of changeover provides us useful insights in understanding the present problems in the Indian context and serves as a guide for putting down an approach for adopting newer technology. He says that though steam reeling yielded substantially higher profits, it did not substantially displace hand-reeling.

The question is as to why the producers did not switch to steam reeling as soon as its superiority had become evident? The imported machinery was efficient, and western markets were eager for good quality silk. The diffusion of steam reeling was to some extent slowed down by the lack of skilled manpower or by the investors' distrust of novelty. Significant as they might have been in the short run, these hindrances could be overcome quite easily by training and by risk-sharing partnerships. So one has to search for some other reason. The usual one is the hostility of the traditional society towards modernisation.

The development of steam-reeling was quite rationally perceived by silk merchants and handloom weavers as a threat to their livelihood, as it could endanger their business or curtail their raw material supply. Hand-reeling production would not have been able to survive so long had it not been for two favourable circumstances. First, low-cost silk was eagerly demanded in Asian markets, where powerloom weaving hardly existed and the low opportunity cost of labour enabled handloom weavers to use an inferior silk and this reflects the Indian situation too well. In Bengal, the traditional charka method was believed to have survived the 'modern' western-type factories just because the silk could be sold in the home market. The low cost of labour coupled with some simple improvements to the machinery made it possible for traditional hand-reeling production to survive and even to thrive for a fairly long period.

Federico emphatically mentions that its final demise could be only delayed but not avoided. The superiority of mechanical reeling was great and growing fast so that in the long run it had to prevail. The most evident consequence of diffusion of steam-reeling was the wholesale adoption of the factory system. The structural reorganisation of the reeling sectors in Italy, Japan and China was a sign of progression from which we can take a lesson or two. Since the 1880s, technical progress had been considered the key to Italian competitiveness; in the opinion of a French expert, innovations 'adopted readily and quickly', have enabled the [Italian] reeling firms to survive and fight on nearly equal terms with their Asian competitors. But better equipment is only a part of the story; the total factor productivity was also higher in Italy than in Japan.

A possible consequence of diffusion of steam was a process of financial concentration - that is, the union of several plants in the same firm. The three main countries showed a remarkable difference from this point of view, even if, as already said, in none of them was the process intense enough to reduce substantially the competition among firms. There is no evidence of concentration in China, possibly because the size of the medium firm was already large enough or else its life span was too short for it to grow. In Italy, a process of concentration accompanied the first stages of diffusion of steam-reeling. Federico cites an official report wherein it is reproachfully stated that, this is a highly divided industry, with an infinite number of small and medium firms and some larger organisations'. By contrast, in Japan, concentration went on until some firms became real giants. The most striking success story was the Katakura company.

Quality Issues

Federico observes that Italian industry was not only the most productive in the world but it also produced the best silk in the world; few sorts could match the Italian quality, none could exceed it. The contribution of the quality of cocoons to this achievement is at best uncertain, whereas that of the machinery and skills of the manpower is on the contrary undeniable..He says that the movement of relative prices can therefore be used as a proxy of quality changes but only if supported by qualitative evidence. Luckily, such evidence is quite abundant because unlike productivity in reeling, quality was a matter of high concern for the buyers of silk.

The Italian industrialists were well aware that 'having lost any hope to compete with the Asian silk on quantity and price', the future of Italian industry hinged on the possibility of maintaining the leadership of quality. When this superiority was vanishing, they seemed not to accept the hard reality, he observes. Federico's observation that the profits of quality improvements for the silk producers were less certain is one of the constraints of quality improvement in the Indian context. He says, on one hand, the task was not easy as it required massive investments in new equipment and in training of manpower and possibly organisational changes. On the other, the returns to all these labour were not immediate (an entrenched bad reputation could be changed only slowly) and by no means sure.

In fact, they depended on the permanence of the quality-premium, which the same improvement put into jeopardy. Therefore, the reluctance to comply with the buyers' requests may well have been rational - provided, of course, that there was a market for low-quality silk. The hard way Japanese silk industry strived to equal and later excel Italian silk industry both in respect of quantity and quality of silk shows the direction in which Indian silk industry can proceed to achieve the desired end results.

Charsley says that the silk thread should be as thin, smooth and regular as practically obtainable. This depends directly on reeling but the ease with which any level of excellence can be attained in reeling depends in turn on the quality of cocoons. This highlights the importance he places on the quality of cocoons. Charsley appears to stress on the aspects of cost of quality and indicates that new improvements are not possible, if the producer cannot realise it in terms of its returns. He says that as far as the financial aspects of efficiency are concerned, increasing the returns to sericulturists does not seem to have been a primary goal in development thinking. When the potential for higher incomes has been stressed, as in connection with the new bivoltine rearing, it has been essentially as a means, with a strong propaganda element, of persuading people to produce a product deemed superior. It has generally been assumed, he says, that a superior product would be more profitable but the industry itself provides at least one clear demonstration that this is not necessarily true; it is at times clearly more profitable to produce charka than basin silk, as the continued health of that."inferior" sector of production bears witness.

Mechanisation demands a homogenous input and American buyers repeatedly complained about the quality of silk sent to their plants. Self interest alone should have dictated that a reform of the Chinese silk industry takes place. Chinese were never noted for a capacity to produce large supplies of any commodity to a uniform quality, and in silk production, which necessarily takes place in small units, the problem was especially troublesome. Reputation, he says, is a widely used means by which quality is ensured.

Unfortunately, as long as we continue with the assumptions of monopolistic competition, requiring many small producers and contestability, requiring easy entry and exit, it is hard to think of any mechanism that will ensure repeated meetings of buyer and seller at low cost and permit reputations to grow, he observes. Regulation is a second alternative. The difficulties with government or professional regulation, however, are just as great if not greater, since the presumption that honest enforcement can be attained is a questionable one.

If the historical evidence is any guide, it will appear that neither of the above approaches is the most effective one. Where quality control has been effectively exercised, it has occurred by enlarging the effective unit of production and by having a self-interested party, the buyer or the seller, assume responsibility. It is well known to economic historians that the Chinese with their dispersed production were unable to provide quality silk while the Japanese succeeded in doing so. The organisation of the industry appears to be the decisive factor. Large units were more prevalent in Japan than in China and the state was willing to give legal sanction to desirable actions.

Quality has a definite tendency to deteriorate in markets with many small producers and easy entry. The most effective check to this determination appears to be the increase of irrevocable fixed assets. When significant amounts of money are invested, the businessman conveys that he plans to stay for sometime to come.

In the long run the only way to stay is by pleasing customers; this requires providing them with the goods they really want, and this long term dependence of producers upon consumers is perhaps the most effective guarantee of quality. Some special welfare problems may also beset the sustainable configuration associated with monopolistic competition. It is plausible that these may involve an undesirably large and an undesirably small set of product varieties. But that is mere conjecture.

Hedonic price model has been built to relate the raw silk price to the levels of its various quality characteristics. Also, the quality attributes of silk yarn together with prices were subjected to factor analysis to study the dimensions of quality. The logit model is used here to illustrate the differentiation between high and low quality silk yarn through weavers" perception.

Cocoons and Silk Markets

Nagaraj et al studied the aspects of spatial integration and price leadership in cocoon markets of Karnataka. The study indicated good integration between the major markets except Kollegal. The dominance of Ramanagaram market has also been assessed. Thomas determined the extent of integration of cocoon and silk yarn markets. The Bangalore silk exchange and Ramanagaram cocoon market which are the biggestmarkets of silk yarn and cocoons in India are studied with a view to determine the efficiency of prices and cointegration of the markets for these products. AREVfA technique has been used for the purpose. Spatial integration and product form integration have been studied. The results suggested that the prices of cocoon and silk yarn are interdependent The study also showed that, though KSMB operations have had an impact on the silk yarn prices, they have tended to become less effective over the years. It is observed that in the case of silk yarn transactions, disequilibrium, in general, occurs when weavers try to wield market power by trying to reduce prices. But since the markets are reasonably efficient, disequilibrium does not persist for long. The changing demand patterns for silk goods both in the export and domestic sectors can influence the production processes through the relatively efficient cocoon and silk markets.

Ram developed regression models for estimating the price of raw silk from the cocoon price. Seasonality components were studied using trend analysis and besides, forecasting cocoon prices and silk prices. In order to counter the effects of abnormal price fluctuations, price stabilisation mechanisms have been suggested. It is reasonable to believe that the operations of markets raises prices generally and dampens their oscillation in same degree. This is not to say that it comes near to eliminating rapid price movement. This remains out of question as long as great seasonal variations in the supply of cocoons and strong movements of the raw silk market occur. Markets improve matters, however, first by making regular and continuous reeling more feasible than exclusively local buying would allow, and secondly by allocating to each market some buying for government filatures. Both these tend to increase the regularity of demand in a field marked generally by irregularity. The government filature or large-scale reeling factories have commitments to their labour forces which require them to maintain more continuous production than any private business does.

Almost continuous buying for them therefore creates a base of demand for good quality cocoons, which tends to bolster prices generally. This has however, to be considered a somewhat extraneous source of advantage and indeed in the longer run an insecure one, since large-scale reeling has already escaped a state of financial crisis since its inception in the 1930s. The existence of cocoon markets does increase the options open to reelers and, as has been seen, there are those who do value the opportunities they provide. Markets allow any reeler more readily to obtain the quantity, quality and price of cocoons he regards as most suited to his own circumstances and the current demand for silk. The market's requirement of cash in hand is sometimes a drawback but there is no evidence that it has so far hindered the growth even of smaller-scale reeling.

Selling by auction in an open market draws attention to the quality of the merchandise in a way avoided by village selling. In the market, batches of cocoons are displayed side by side, making, comparison inevitable and perceived quality is directly reflected in prices varying from lot to lot.

Naik. G & Babu have analysed the demand and supply prospects for high quality raw silk in India. The supply of and demand for quality raw silk have been estimated by using micro-economic models. Certain estimates have been used where ever data was not available. They have estimated that the total high quality silk production could meet at the most 60 percent of the estimated demand. They have proposed certain policy options into demand related and supply related aspects. They have cautioned about the negative implications the import of Chinese raw silk has on the development of Indian silk industry. The diversion of imported silk into domestic sector is benefiting only the consumers. By avoiding this diversion, the demand for domestically produced high quality silk would have increased for the good health of the industry. Some of their recommendations include - improvement in research and extension facility, adequate supply of inputs and proper marketing facilities and encouraging modification in the production system.

Federico has explained in detail the market for silk and the efficiency and risk in the modern institutions. The independence from the merchants and the keen competition among intermediaries were not the only advantages of the modern institution from the producers' point of view. The European market places were remarkably well organised. Since the early 19th century offered facilities such as the conditioning of silk, and a set of standard market regulations, the transaction costs were reduced. Modern institutions also offered traders, abundant information at the price of a subscription to the specialised press.

Levich tries to answer, whether traders make the best use of the market information and whether the silk market can be called efficient if 'prices fully reflect all the available information' so that investors [dealers] cannot earn 'unusual profits' by exploiting available information'. One could use past prices, discovering and using whatever statistical pattern exists in them. So, in an efficient market, the price movements should not exhibit any detectable pattern: they should remain steady because all information is used and move (unpredictably) only when new data are made available. In statistical jargon, they should be random walks. In an efficient market the difference between the predicted and the actual future prices should not be correlated to any of the data available to dealers at the moment of forecasting.

Federico observes that the most frequent proposal was to set up a permanent marketing board, either as a private enterprise (as a joint-stock trading company or as a co-operative among industrialists, funded with a fixed contribution on the silk sales), or as an official agency. According to one of these proposals, the board should act as a market maker, intervening with purchases, sales or any other suitable measure, to contrast any excessive price movements - both on the rise and on the fall. The efforts in this direction proved that no private company could ever raise the capital necessary to stabilise the world silk prices. Thus, the hopeful industrialists turned to the state, so resuming a very old tradition.

Attempts to set up marketing boards failed because stabilisation was a long-term solution to a short-term problem namely, speculative losses. A permanent stabilisation of prices, if possible (which is highly doubtful), would have stripped the industrialists of the opportunity to make speculative gains.

What the industrialists really wanted was a syndicate that cornered the market to raise prices when they were falling too much. From the 1910s onwards, the Italian and Japanese governments substituted private speculators in organising corners - almost inevitably with very poor results. These failures are hardly surprising. A corner could create an artificial scarcity of silk and compel the weavers to buy at a higher price, but the syndicate sooner or later had to sell the silk it had amassed, causing prices to fall again. Risk remained an unavoidable feature of the silk industry.

Federico observes that there are a large number of possible sources of competitiveness - structural differences in the location of reeling and/or in the inherent risk of the business, product differentiation in manufacturing, the technology, the availability of low-cost inputs, the efficiency of markets for goods and for credit and the state support. The technical progress was indeed very important but it was largely an outcome of market competition, i.e. of a spontaneous drive from below rather than of a coordinated effort from above. The advantages of size were rather very dubious in a perfectly competitive market. The trading companies were probably an efficient response to market failures, but they were defenseless in the face of market vagaries.

Charsley provides an account of the working of the old silk market in Karnataka, when the Karnataka silkworm seed, cocoon distribution and control amendment act passed in September 1979 required the producers of raw silk to sell only through an official exchange established in Bangalore. Such an account would provide a basis for evaluating the reforms. The author provides a detailed account of the ways and means of silk transaction and the overwhelming presence of the traders in the whole transaction. The author says that the reelers vary widely in status also, according to the scale and style of their operations and the quality of the silk they produce. The author observes, that at the top end of the market, there are reelers who belong to the landed rural elite with large capital investments in buildings, machinery and stock, and very often with members of their families in graduate professions. At the bottom end of the market status, greater differences exist, contributing to the niche available for brokers. He says that it is important to realise that reelers are necessarily drawn from the more ambitious, determined and successful in the communities from which they came and that there is no debt relationship to express and maintain any clear subordination of reeler to merchant. The author observes that there was no evidence of any serious local dissatisfaction with the private raw silk market the industry had developed.

Charsley concludes by saying that the difficulty of rivaling the private market system built up over the years was obvious. The business, he says, is inevitably complex and varied, since the types of silk, its variable quality, and the wide range of the circumstances and modes of operation of the reelers make it unlikely that any simple scheme will fit all the needs that impinge upon it. The strength of the diversified private market is that it has been constructed by individuals responding to real needs and their corresponding opportunities. The middleman does exacts his price for the service provided, but it must be doubtful whether this has amounted in the silk industry to any sort of 'economic stranglehold'. The author expresses his skepticism, saying that it remains to be seen whether as satisfactory a service can be provided by public agents who, after all, exact their price too. His apprehensions have been proved to be right beyond doubt.

Now, after 21 years of its existence, the silk exchanges have not been able to make any impact on the system of silk marketing in Karnataka. The private trade continues to operate with renewed strength with two important associated ills. The onus of maintaining the infrastructure established rests on the tax payers and it has created an avenue for fleecing the reelers and weavers in the name of law.

Baurer observes that officials with extensive powers over traders and producers are in a position to abuse their powers for their own benefits. This, he says unlike the abuses of the traders, cannot be competed away and redress from higher authorities may not be practicable. This aspect of market failure has been brought out by Charsley, in his analysis of the impact of the series of market interventions by the government.

The proposition that improvement in markets cannot be secured by official intervention has been put forward by Baurer, on the basis of experience of a multitude of efforts in different countries, which has been examined in the context of silk industry in Karnataka by Charsley. The functions performed by the traders or the middlemen in the silk industry have been examined. He says, that there are several hundreds of them, found mainly in Bangalore. With the establishment of the silk exchange, only a few local twisters and weavers were in a position to come to auctions and it became immediately obvious that without the traders there would be insufficient buyers for there to be a market at all. The general effect was to push the entire silk market into an ever-increasing illegality and even invisibility.

Discussing about the subtler issue concerning the maintenance of quality, silk produced in Indian conditions is a variable, hand-worked commodity. The quality of any particular lot is established conclusively when it is used, first in twisting and then for weaving. The way merchants operated in the past allowed a feedback on quality, which is not the case in the present market situation and this has led to deterioration in quality. Referring to the success of cocoon markets, he says, for reelers they provided a more regular secure source of raw material in the quantities and of qualities required. For the farmers, the cocoon markets were thought to be generally effective in securing good prices.

Support to Silk Reeling

In the traditional institutions, financing was often a part of the transaction in silk as merchants anticipated the proceedings of the final sale to go to weavers. In most cases, the same funds were lent several times, in a sort of downstream flow. Sometimes, the lenders at the beginning of this sort of chain used their own money, but, more frequently they borrowed from banks. Indirect financing reduced the banks' risks because they could use the merchants personal properties as a further collateral and because merchants, as insiders, could screen applicants more effectively. The intermediaries gained the interest rate differential.

Institutions were indeed important but hardly decisive according to Federico. The total transaction costs might have been much higher than the quoted official fees if imperfect competition had given the intermediaries the opportunity to gain extra profits at the producers" expense. This was surely not the case with modern institutions. On the contrary, many descriptions of traditional institutions stress the existence of market imperfections, especially concerning the power of buyers to set the price.

There is little doubt hat the sellers received less than the market price for their product. Neither interlinkage nor a reduced price is in itself, an evidence of oligopsony or exploitation. Interlinkage is basically a loan on pledge of future production. The final price of the good has to include the interest on loan, the risk premium and the remuneration for whatever service the buyer provides. In a competitive market, the buyer could charge only other services - and interlinkage does not entail any oligopsonistic clout.

Charsely says that there exist a substantial section of reelers in Karnataka, who do regard the cocoon market as advantageous and regularly and voluntarily use it. Whatever the quality, quantity and price level the reeler seeks, the market is the place where he is most likely to be able to meet his requirements. Many of those for whom credit is the major consideration, the market however, has the fundamental drawback of normally requiring cash in full and on the spot Charsley highlights the importance of capital and says that finance also appears as an issue in developmental context.

In recent years there has been a widespread official assumption that the major limitation on progress in the industry has been the availability of capital. In the case of sericulture, study of the way production actually works shows that, were reelers to be successfully prevented from obtaining cocoons directly from their producers in the villages, as has been the intention, this would destroy the financial self-sufficiency of the industry, at present considerable, and make reeling activity heavily dependent upon funds being available from the public purse. This would indeed constitute a transformation but hardly one which many would regard as a desirable improvement.

A moderate policy of support was not a sufficient condition for the growth of silk production if the factor endowment was not suited to it. Sericulture could not prosper where labour was, or was becoming expensive. Of course, this failure does not rule out the possibility that the development of sericulture could have been forced by extreme means such as a very large subsidy or (in consuming countries) a prohibitive duty. This policy would have been very expensive, both for the consumers and the state coffers. At any rate, no country resorted to it. Was state support a necessary condition for success in suitable countries? Many people believed so.

The concern with finance, on other extension propaganda grounds, might be expected to arise and has indeed arisen in two connections. The high relative cost of Indian silk, first has been a preoccupation at the highest planning levels. It has been seen as necessitating protection in the home market and as hindering the export potential. Such concern was evinced in both 3rd and 4th five year plans. It might be thought that this would have inevitably led attention towards commercial considerations. In fact the high cost was attributed to technical factors, suggesting no need, that is to say, for close attention to the realities of commercial, i.e. money-making production.

From the point of view of silk reeling, it is necessary to study the reeling industry at a micro level to grasp the real problems of the individual reeler and the literature does not reflect such attempts. A broader study of the total reeling sector as attempted by some of the authors would not yield this understanding. In order to study silk reeling at the micro level i.e., a typical reeling unit, two pre-requisites are to be met. The first is a clear measure of cocoon quality. Second is the establishment of the relationships among cocoon price, quality and quantity, as also between cocoon price and raw silk price.

Bibliography

Anonymous, *History of Silk,* 2003.

Chen, Yao-Min, "Qing Mu Shunde Jiqi Cao Si Ye De Cai Sheng, Fa Zhan Jiqi Ying Xiang" *(The Beginning, Development and Significance of Machine-Reeling Industry in Late-Qing Shunde)* in Ming Qing Guangdong SheHui ChinJi XingTai YanJiu (Studies on the Socio-economic Conditions of Ming and Qing Guangdong) edited by the Historical Association of Guangdong,

Chen, Zi-yu, Jingdai Zhongguo de Jijie Saosi Gongye, 1860-1945, (*The Silk Industry of Modern China*, 1860-1945). Taipei, Taiwan:Institute of Modern History, Academia Sinica, 1989.

Furuta, Kazuko, "Technology Transfer and Local Adaptation: the Case of Silk-Reeling in Modern East Asia", *A Dissertation* submitted to Princeton University, 1988.

Moedoko, W., "Bombyx Mori Silk recent Development and Prospect in Indonesia", *A Paper to presentation Fourth International Conference on Wild Silk Moth*, Yogyakarta, Indonesia, 2002.

Nakabayashi, Masaki, "Kindai Shihon Shugi No Soushiki (*Organization in Modern Capitalist: Governance of Trade and Production in the Silk Reeling Industry*)", University of Tokyo Press, 2003.

Segi, Hideyas, "A Brief History of *Italian Style Reeling Machines in Journal of Okaya Silk Musuem*, No. 3. Published by Okaya Silk Musuem, 1998

So, Alvin Y., "The South China Silk District: Local Historical Transformation and World-System Theory", Albany: State University of New York Press, 1986.

Soda, Saburo, Chugoku Kindai Seishi-gyo-shi No Kenkyu (*Study on the Modern Chinese Silk Industry*). Tokyo: Kyuko Shion, 1994.

World Bank, "Bangladesh-Silk Development Pilot Project", 2001.

Xu, Xin-wu (Ed.), Zhongguo Jindai Saosi Gongyeshi (*Modern History of Chinese Silk-reeling Industry*). Shanghai: People's Publishing House, 1990.

Zanier, Claudio, "Where the Roads Meet, East and West in the Silk Production Processes (17th to 19th Century)", Kyoto: Instituto Italiano di Cultura Scuola di Studi Sull' Asia Orientale, 1994.

Index